DATE DUE

JAN 2 5 1982		
FEB 1 7 1982		
RETURNED		
11·8·89		
NOV 1989		
RETURNED		
AUG 2 9 1996		
	ʼ	
RETURNED		
MAR 1 7 2000		
MAY 2 4 1999		

DEMCO NO. 38-298

**Predicting the Properties
of Mixtures:** Mixture Rules
in Science and Engineering

Predicting the Properties of Mixtures: Mixture Rules in Science and Engineering

Lawrence E. Nielsen
Monsanto Company
St. Louis, Missouri

Marcel Dekker, Inc.   New York and Basel

Library of Congress Cataloging in Publication Data

Nielsen, Lawrence E
    Predicting the properties of mixtures.

    Includes bibliographies and indexes.
    1. Mixtures. I. Title.
QD541.N46   660.2'9   77-16705
ISBN 0-8247-6690-3

MARCEL DEKKER, INC.

270 Madison Avenue, New York, New York 10016

Current printing (last digit):
10 9 8 7 6 5 4 3 2 1

PRINTED IN THE UNITED STATES OF AMERICA

To the memory of
Hans C. Nielsen

PREFACE

Scientists and engineers very often need to know the properties of mixtures. Although much information on the properties of mixtures is scattered throughout the literature, generally the discussions are limited to only one specific type of system or property. Often the articles in the literature describe attempts to make calculations of properties based upon some theory. Little effort has been made on how to best empirically estimate the properties of all kinds of mixtures (both one-phase and two-phase) from a very general point of view. This author is not aware of any textbook or monograph which adequately deals with the subject of how to best calculate the properties of various kinds of mixtures with a minimum amount of information derived either from theory or from experiments. It is the purpose of this small monograph to fulfill this need to have the proper equations and guidelines for predicting the properties of all kinds of mixtures.

Mixtures include one-phase miscible systems and two-phase systems such as composite materials and suspensions. However, all binary mixtures can be divided into three general classes: 1. One-phase miscible mixtures. 2. Two-phase systems with one continuous phase and one dispersed phase. 3. Two-phase systems with two continuous phases. There are general mixture rules or equations for each of these classes of mixtures, and one must select the proper type of equation for accurate estimation of any given property. One also must know what information is required in addition to the corresponding property of the components and their concentrations in order to make valid estimates of the property of the mixture.

In this monograph, emphasis is placed on the proper selection of a mixture rule for a given system and property. In some cases theory can be used to estimate the factors in addition to concentration that are required to make accurate predictions. However, so little theory is available in most cases that empirical methods must be used to evaluate these additional factors. These additional factors change with the general class of mixture being considered. Many examples are taken from the literature to illustrate how to estimate various properties for each of the different classes of mixtures.

The author thanks Don Carter, Eli Perry, and Murray Underwood for reading the manuscript and offering many suggestions for improving the text. Bobbie Kaplan typed the manuscript. My wife, Deanne, proofread the manuscript and helped with the indexes.

Lawrence E. Nielsen

CONTENTS

Predicting the Properties
of Mixtures: Mixture Rules
in Science and Engineering

Chapter 1

INTRODUCTION TO MIXTURES AND MIXTURE RULES

I.   INTRODUCTION

Mixtures of two or more components are found in countless cases in everyday life. Scientists and engineers constantly need to know the properties of these mixtures. It would be desirable to be able to predict easily the properties of the mixtures from just the corresponding properties of the components and their concentrations. However, in general, additional information must be known about the nature of the mixture if accurate prediction of properties is to be made. What kinds of information in addition to concentration and properties of the components are required in order to predict properties of mixtures? The additional factors which are needed to characterize mixtures include interactions between the constituents, particle size and shape, and the nature of the packing found in the mixture.

1

Hundreds of equations (both theoretical and empirical) can be found in the literature for predicting the properties of mixtures. Many of these equations can be shown to be derivable from only a few simple, but very general, mixture rules. The main objective of this monograph will be to present these general mixture rules, to show the type of mixture to which each of the rules is applicable, and to show how these rules may be used to predict properties of mixtures in practical applications.

II. CLASSES OF MIXTURES

Mixtures can be divided into two broad classes: 1. Single phase systems in which the components are completely miscible or soluble in each other. 2. Two-phase or multi-phase systems in which the components are insoluble (or only partially soluble) in each other. In the cases where the constituents are soluble in one another to give a one-phase system, the interactions between the molecules and how they pack are important in determining the properties of the mixture. In the cases where the components are not soluble in one another but form two-phase systems, the factors which must be considered in the prediction of properties are: 1. Which phases are continuous and which are dispersed? 2. What is the shape of the particles, and what is the morphology of the system? 3. How do the particles pack? 4. If the particles making up a phase are not spherical, how are they oriented? 5. What kind of interaction occurs at the interfaces?

What are some of the properties of mixtures which one might want to predict? Table 1 lists some of these properties for one-phase systems while Table 2 lists some of these properties for two-phase or multi-phase systems.

Table 1

Examples of Properties Predictable for One-Phase Systems

1. Density (or specific volumes) of liquid mixtures
2. Refractive index of liquid mixtures
3. Dielectric constant of mixtures
4. Viscosity of liquid mixtures
5. Boiling point diagrams
6. Thermal conductivity of liquid mixtures
7. Critical temperatures and pressures of mixtures
8. Surface tensions of liquid mixtures
9. Heat capacity of mixtures
10. Glass transition temperatures of copolymers
11. Glass transition temperatures of plasticized polymers
12. Glass transition temperatures of compatible polymer mixtures
13. Thermodynamic properties

Table 2

Examples of Properties Predictable for
Two-Phase Systems

1.   Elastic moduli of composite materials
2.   Electrical conductivity of mixtures and composite
     materials
3.   Thermal conductivity of mixtures and composite
     materials
4.   Dielectric constant of mixtures
5.   Gas and liquid permeability through composite
     materials
6.   Diffusion through mixtures
7.   Viscosity of suspensions
8.   Flow through porous media
9.   Coefficient of thermal expansion of mixtures

III. THE SIMPLEST MIXTURE RULES

The simplest mixture rule is often called "the rule of mixtures." It is

$$P = P_1\phi_1 + P_2\phi_2 . \tag{1}$$

The given property of the mixture is P; P_1 and P_2 are the corresponding property of the components 1 and 2 of the mixture. The concentrations of the components are ϕ_1 and ϕ_2. The concentration terms may be weight fractions, volume fractions, or mole fractions depending upon the kind of property being measured. In this work ϕ shall represent any of the concentration terms in general, but in most specific cases, ϕ shall be the volume fraction. Equation 1 indicates that the property of a mixture is just a linear combination of the properties of the components making up the mixture, and it is a good representation of the experimental facts in some cases.

Another simple equation is the "inverse rule of mixtures":

$$\frac{1}{P} = \frac{\phi_1}{P_1} + \frac{\phi_2}{P_2} \tag{2}$$

This equation also is valid for some mixtures. Equations 1 and 2 require that one only needs to know the value of the given property for each of the components and their concentrations. It should be remembered that the concentrations are not independent of one another but are related by the equation

$$\phi_1 + \phi_2 = 1. \tag{3}$$

In a great many cases, equation 1 gives the highest value
that property P of a mixture can have while equation 2
gives the lowest value that P can have. Thus, equation
1 often is called the upper bound of a property, and
equation 2 is called the lower bound.

Figure 1 illustrates an example of cases where equations 1 and 2 apply. Suppose two materials are glued together and then stretched by a force in the direction of
the arrows. In this case P can be the elastic modulus
of the material, i.e., the resistance to stretching of
the composite to a unit force. In Figure 1, ϕ_1 and ϕ_2
are the volume fractions (or areas) of materials 1 and
2. The bottom section of the figure is a plot of the
property (modulus in this case) as a function of concentration of component 2. In one graph, a linear scale is
used for P, and in the other case a logarithmic scale is
used for P. In the linear case, equation 1 gives a
straight line, but equation 2 gives a curved line. When
a logarithmic scale is used for P, both cases A and B
give curved lines, but the lines are symmetrical about a
straight line joining the end points. The property P
could be any one of many properties, not just elastic
modulus. For example, P could be electrical or thermal
conductivity, diffusion coefficient, liquid or gas permeability, and the transmission of light through the composite systems.

IV. MORE GENERAL MIXTURE RULES

In the great majority of mixtures, the simplest mixture rules are not capable of accurately predicting the
properties of a binary mixture. To make accurate predictions, some kinds of additional information must be
available, and this additional information must be applied
to an equation appropriate to the type of mixture. Three
very general types of mixture equations will be considered.

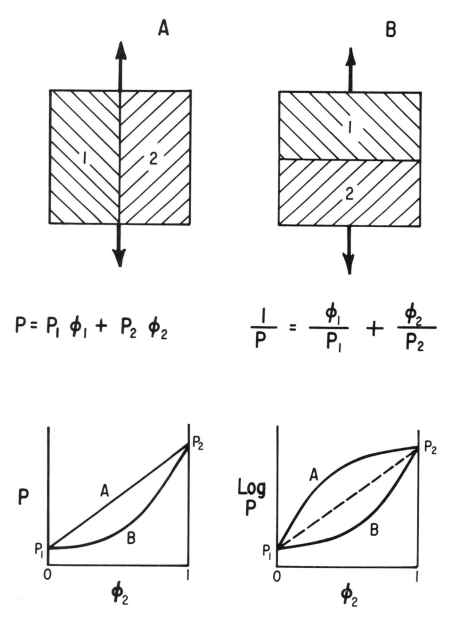

Figure 1. An example of the application of "the rule of mixtures" and "the inverse rule of mixtures" for predicting such properties as the elastic modulus of simple composite systems. Note the difference in symmetry of the curves for linear P plots compared to log P plots.

For one-phase systems (1,2):

$$P = P_1\phi_1 + P_2\phi_2 + I\phi_1\phi_2 \qquad (4)$$

I is an interaction term which can be either positive or
negative. When two substances are soluble in one another,
I is related to such factors as intermolecular interactions or the way in which the molecules pack in the
mixture.

Another class of mixtures, including suspensions, has
a continuous phase and a dispersed or discontinuous phase.
The properties of such two-phase systems often can be
predicted by the following general mixture rule (3-7):

$$\frac{P}{P_1} = \frac{1 + AB\phi_2}{1 - B\psi\phi_2} \qquad (5)$$

where A is a constant which depends upon the shape and
orientation of the dispersed particles and upon the
nature of the interface. It can vary between zero and
infinity. The constant B depends upon the value of the
properties of the components and is defined by

$$B = \frac{P_2/P_1 - 1}{P_2/P_1 + A} \qquad (6)$$

In this type of representation of such a mixture, P_1
is the given property of the continuous or matrix phase,
and P_2 is the property of the discontinuous or dispersed
phase. The factor ψ is a reduced concentration term
which is a function of the maximum packing fraction ϕ_m.
Since it is impossible in general to pack particles so
that there is no void space between them, particles
appear to occupy a volume which is greater than their
true volume. The maximum packing fraction is defined as

$$\phi_m = \frac{\text{True volume of the particles}}{\text{Apparent volume of the particles}} \qquad (7)$$

or

$$\phi_m = \frac{\text{Bulk density}}{\text{True density of the particles}} \cdot \qquad (8)$$

The factor ψ can be defined in many ways, but the simplest equation is (8):

$$\psi \doteq 1 + \left(\frac{1 - \phi_m}{\phi_m^2} \right) \phi_2 \cdot \qquad (9)$$

When $\phi_m = 1$, $\psi = 1$ at all concentrations. The concentration of the dispersed phase is ϕ_2. It will be shown in Chapter 3 that several kinds of mixture equations can be put into the form of equation 5.

The third type of mixture consists of two separate or discrete continuous phases (3, 9-11). Examples of such systems are interpenetrating networks, mats and felts filled with air or some other material, connected open-celled foams filled with a gas, liquid, or solid, and laminates consisting of sheets of two or more materials glued together. The properties of such systems often can be predicted by the following equation:

$$P^n = P_1^n \phi_1 + P_2^n \phi_2 \quad ; \quad -1 \leq n \leq +1. \qquad (10)$$

The constant n depends upon the type of system and the kind of property being studied. Equation 10 has a type of symmetry not found inherent in equation 5. Since both phases are continuous, the equation is symmetrical with respect to both components 1 and 2, and either phase can be designated as component 2.

A special case of equation 10 is found when $n = 0$. The resulting equation is known as the logarithmic rule of mixtures. After some mathematical manipulation of equation 10, the resulting equation when $n = 0$ is:

$$\log P = \phi_1 \log P_1 + \phi_2 \log P_2 . \tag{11}$$

It can be shown that the simplest rule of mixtures is obtained when $A = \infty$ in equation 5 and when $n = +1$ in equation 10. The inverse rule of mixtures results when $A = 0$ in equation 5 and when $n = -1$ in equation 10. Thus, both equations 5 and 10 can cover the complete range of properties between the upper and lower bonds as defined by equations 1 and 2. However, equations 5 and 10 do not cover the total possible range of properties in the same manner. This difference in the shape of the curves is illustrated in Figure 2. On a plot of $\log P/P_1$ versus ϕ_2, equation 10 gives a straight line when $n = 0$. The closest match to this straight line when equation 5 is used is to let $A = (P_2/P_1)^{1/2}$. Then, equations 5 and 10 are the same at the end points and at the middle, but the two equations are not the same at other concentrations. The general trend always is for equation 5 to give higher values of P/P_1 at concentrations of ϕ_2 between zero and 0.5. At concentrations above $\phi_2 = 0.5$, equation 10 gives larger values of P than does equation 5. This same point is made in Figure 3 where equation 5 is plotted for various values of A between zero and infinity. Superimposed on this figure as a broken line is equation 10 with $n = 0$. All values of n greater than zero would give curves above the broken line, and all values of n less than zero would give curves below the broken line.

V.   CONCENTRATION VARIABLES

The concentration of the components in a mixture can be expressed in many ways. Generally, however, there is only one correct (but often unknown) way of expressing the concentration. This "correct" concentration depends upon the nature of the property of interest and on the kind of mixture. The most common concentration terms used in mixture rules are weight fraction, volume fraction, and mole fraction. In most cases there is no theoretical reason for using weight fraction, but it is often used because of convenience. Mole fraction usually is the proper concentration term in many cases when the two components are soluble in one another, and when the molecules interact with one another on a number basis. In some cases the volume of the molecules is more important than their number; then volume fraction is the correct concentration to use. In two-phase systems, the components do not interact with each other on a molecular basis. The volume of the components is the important factor in two-phase systems, so the concentration should usually be expressed in terms of volume fractions. The proper use of a mixture rule requires that the correct concentration variable be used; unfortunately, the proper choice of concentration units is not always obvious.

The weight fraction W_1 of component 1, the volume fraction ϕ_1 of component 1, and the mole fraction X_1 of component 1 are defined by the following equations for two component mixtures:

$$W_1 = \frac{w_1}{w_1 + w_2} \qquad (12)$$

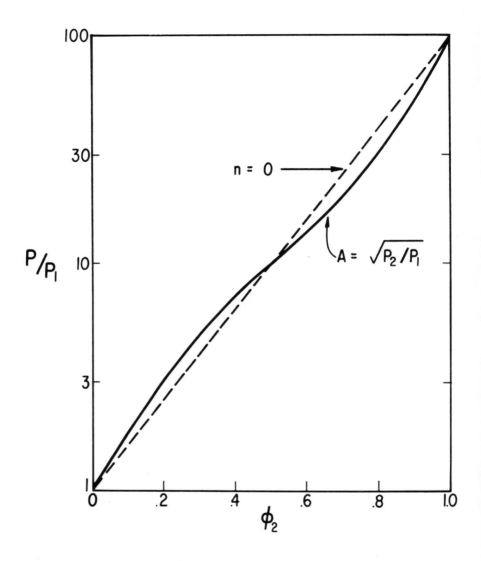

Figure 2.   Comparison of the mixture equations for the two types of two-phase systems. The solid line is from equation 5 with $\psi = 1$, and the broken line is equation 11 or equation 10 with $n = 0$. $P_2/P_1 = 100$.

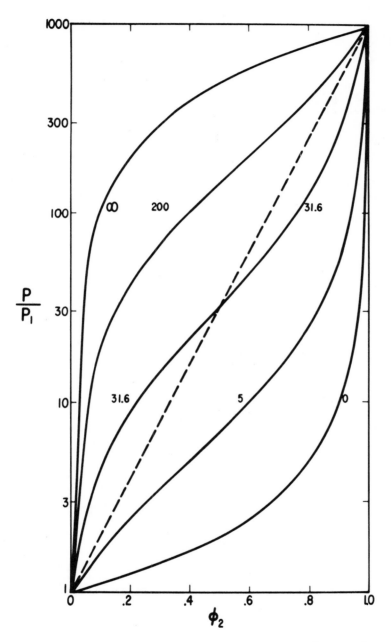

Figure 3.   Equation 5 for various values of A.   The
broken line is equation 10.

$$\phi_1 = \frac{V_1}{V_1 + V_2} = \frac{w_1/\rho_1}{w_1/\rho_1 + w_2/\rho_2} \tag{13}$$

$$X_1 = \frac{n_1}{n_1 + n_2} = \frac{w_1/M_1}{w_1/M_1 + w_2/M_2} \tag{14}$$

In these equations, w_1 and w_2 are the weights of materials 1 and 2 in the mixture, V_1 and V_2 are their volumes, and n_1 and n_2 are the number of moles of the components. The densities of the components are ρ_1 and ρ_2, and their molecular weights are M_1 and M_2.

Useful interrelationships are:

$$\frac{1}{\phi_1} = 1 + \frac{w_2}{w_1}\left(\frac{\rho_1}{\rho_2}\right) = \frac{1}{W_1} - \frac{w_2}{w_1}(1 - \rho_1/\rho_2) \tag{15}$$

$$\frac{1}{X_1} = 1 + \frac{w_2}{w_1}\left(\frac{M_1}{M_2}\right) = \frac{1}{\phi_1} - \frac{w_2}{w_1}\left(\frac{\rho_1}{\rho_2} - \frac{M_1}{M_2}\right) \tag{16}$$

$$\frac{1}{X_1} = \frac{1}{W_1} - \frac{w_2}{w_1}\left(1 - \frac{M_1}{M_2}\right) \tag{17}$$

Corresponding equations for W_2, ϕ_2, and X_2 can be written by switching the subscripts 1 and 2 in the above equations.

The rule of mixtures can be written in more general forms than that given in equation 1 (12):

$$P = \frac{P_1}{1 + K_1(n_2/n_1)} + \frac{P_2}{1 + K_2(n_1/n_2)} \tag{18}$$

where K_1 and K_2 are constants.

If $K_1 = K_2 = 1$, then $P = P_1 X_1 + P_2 X_2$.
If $K_1 = M_2/M_1$ and $K_2 = M_1/M_2$, then $P = P_1 W_1 + P_2 W_2$ since
$w_1 = n_1 M_1$.

If $K_1 = \dfrac{M_2 \rho_1}{M_1 \rho_2}$ and $K_2 = \dfrac{M_1 \rho_2}{M_2 \rho_1}$, then $P = P_1 \phi_1 + P_2 \phi_2$.

Or in still another form:

$$P = \frac{P_1}{1 + k_1 (w_2/w_1)} + \frac{P_2}{1 + k_2 (w_1/w_2)} . \qquad (19)$$

If $k_1 = k_2 = 1$, then $P = P_1 W_1 + P_2 W_2$.
If $k_1 = M_1/M_2$ and $k_2 = M_2/M_1$, then $P = P_1 X_1 + P_2 X_2$.
If $k_1 = \rho_1/\rho_2$ and $k_2 = \rho_2/\rho_1$, then $P = P_1 \phi_1 + P_2 \phi_2$.

For other values of K_1 and K_2 in equation 18 or
for other values of k_1 and k_2 in equation 19, the
concentration scales are distorted. Much data can be
force fitted to equations 18 or 19 by the proper choice
of the constants which are used to distort the concentration scales. The use of such distorted concentration
scales generally is pointless and should be avoided unless there is some theoretical justification for their
use. Generally, the use of distorted concentration
scales can be avoided by using mixture equations such as
equation 4 where only a single interaction term is needed.

Reduced concentrations are useful as in equations
5 and 9 where it is physically impossible to completely
cover the entire range between the two components.

The concentrations are not related linearly, so the
direct substitution of one kind of concentration for
another will give a distorted concentration scale in the
mixture equation used to predict a property. As a result,
the correct value of property P will not be predicted

when the incorrect type of concentration variable is
used. This is illustrated in Figure 4. In this example,
M_1 = 100, M_2 = 150, ρ_1 = 1.0, and ρ_2 = 0.80. A weight
fraction of 0.4 gives a mole fraction of about 0.5 and a
volume fraction of about 0.35. Thus, it is obvious that
using a weight fraction of 0.4 in an equation in which
the proper concentration variable is mole fractions leads
to large errors in predicting the desired property.

VI. WHY MIXTURE RULES MAY BE INVALID

There are some properties which generally cannot be
expected to be predicted by mixture rules. There are
other cases in which a mixture equation would be expected
to hold, but something unexpected changes as a function
of concentration, or some unexpected other phenomenon
occurs which makes the equation invalid. The following
are examples of situations in which mixture rules may be
invalid:

1. The nature of the system changes in some manner
as a function of composition that is not taken into
account by the assumptions used to derive the equation.
For example, the morphology of a composite system may
change with composition such that spherical particles
change to rod-like particles, or phase inversion may occur.

2. The nature of one or more of the constituents may
be changed by the presence of the other, and this change
is not taken into account. For example, the electrical
conductivity of an aqueous salt solution cannot be calculated from the electrical conductivity of pure water
and salt.

3. Another property, which affects the measured
property, simultaneously changes nonlinearly with the
measured property. An example might be the viscosity of
a mixture of liquids in which the density changes in a
nonlinear manner as a function of composition.

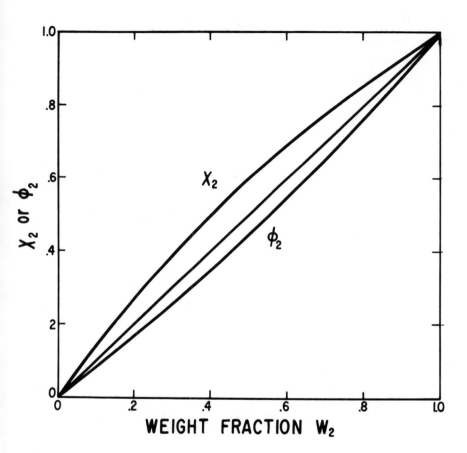

Figure 4.   Mole fraction X_2 or volume fraction ϕ_2 plotted
            as a function of weight fraction W_2 for the
            cases where $M_1 = 100$, $M_2 = 150$, $\rho_1 = 1.00$,
            and $\rho_2 = 0.8$.   The center line is a straight
            line joining the points 0 and 1 on the
            concentration scales.

4. Some properties of mixtures may depend less upon
the corresponding property of the constituents than upon
some other property or characteristic of the mixture.
For example, the strength of a composite material may
depend very strongly upon the nature of the interface
between the materials and upon the shape of the particles
in the composite. The nature of stress concentrators
has a tremendous effect upon the strength of composite
materials. The impact strength of rubber filled brittle
plastics is largely determined by the strength of the
adhesive bond between the rubber and plastic phases.

5. As pointed out earlier in this chapter, the
proper mixture rule must be used. There are many examples where an attempt was made to use a mixture rule
in a situation in which it was not intended. Of course,
the predictions generally are very inaccurate in such
situations. For example, one should not use equation 5
when equation 10 should have been used.

VII. REFERENCES

1.   J. R. Partington, Advanced Treatise on Physical
      Chemistry, Vol. 2, pp 23, 31, Longmans Green,
      London, 1951.

2.   J. R. Partington, Advanced Treatise on Physical
      Chemistry, Vol. 4, p. 73, Longmans Green, London,
      1953.

3.   L. E. Nielsen, Mechanical Properties of Polymers
      and Composites, Vol. 2, Marcel Dekker, New York,
      1974.

4.   L. E. Nielsen, Chemtech., 4, 486 (1974).

5.   L. E. Nielsen, J. Appl. Phys., 41 4626 (1970).

6.   E. H. Kerner, Proc. Phys. Soc. (London), B69, 808
      (1956).

7.   J. C. Halpin, J. Compos. Mater., 3, 732 (1969).

8.   T. B. Lewis and L. E. Nielsen, J. Appl. Polymer
      Sci., 14, 1449 (1970).

9.   W. E. A. Davies, J. Phys., D4, 1176, 1325 (1971).

10.   G. Allen, M. J. Bowden, S. M. Todd, D. J. Blundell,
      G. M. Jeffs, and W. E. A. Davies, Polymer, 15, 28
      (1974).

11.   L. E. Nielsen, J. Appl. Polymer Sci., 19, 1485
      (1975).

12.   L. T. Carmichael and B. H. Sage, AIChE J., 12,
      559 (1966).

Chapter 2
ONE-PHASE BINARY MIXTURES

I.   INTRODUCTION

If two materials form a one-phase system, this im-
plies that the two materials are soluble in one another.
Thermodynamically, there is always an entropy of mixing
which tends to make two substances miscible in one another.   In many cases there also is an energy-of-mixing
which can greatly influence the solubility behavior and
thereby many other properties as well.   This energy-of-
mixing term is the result of some kind of an interaction
between the molecules making up the mixture.   Interaction
effects also may arise from changes in how molecules pack
due to preferred orientations.   This type of interaction
can be largely an entropy effect rather than an energy-of-
mixing effect.

One-phase systems can be gases, liquids, or solids.
In general, however, the mixing must occur in either the
gaseous or liquid state.   For a given composition, the
degree of mixing could be very different for different
systems.   The molecules can be completely random, they

can exist in an associated state as clusters, or the two
kinds of molecules can exist in various types of com-
plexes. One would expect that the properties of these
various kinds of mixtures would be different. These
differences in properties would show up as differences
in the interaction terms of mixture rules. However,
except in a few cases, our theories and experimental
techniques are not sophisticated enough to enable us to
predict theoretically these interaction terms from the
degree or type of mixing that exists in a specific system.
Therefore, in most cases, the interaction terms must be
determined empirically from experimental data.

Although mixture equations can be developed for
systems containing more than two components, the general
equations rapidly become very complex, so we shall limit
ourselves to binary systems. Some of the types of properties which can be estimated for one phase mixtures
were given in Table 1 of Chapter 1. Many other properties could be added to this list.

II. THE GENERAL EQUATION FOR ONE-PHASE MIXTURES

As pointed out in Chapter 1, a useful general equation for predicting properties of one-phase binary mixtures is [1-3]:

$$P = P_1\phi_1 + P_2\phi_2 + I\phi_1\phi_2 . \qquad (1)$$

P is the value of the property of interest for the mixture, and P_1 and P_2 are the corresponding values of the
property for materials 1 and 2, respectively. The concentrations of materials 1 and 2 are ϕ_1 and ϕ_2. For one-
phase systems, ϕ_1 and ϕ_2 are very often the mole fractions of the components, but in some cases they may be
volume fractions or even weight fractions. Thus, ϕ is a

general concentration fraction. When specific systems
are discussed, ϕ will also represent volume fractions, X
will be mole fraction, and W will be weight fraction.
The interaction term I can be either positive or negative,
depending upon the system. The new factor in a mixture
which did not exist in the pure components is the interaction between the two different kinds of molecules.
Thus, the effect of the interaction term on the given
property also depends upon the concentration of each of
the two kinds of molecules.

Other general mixture equations could be used for
one-phase mixtures. However, it can be shown that many
of these equations can be rearranged to the form shown
in equation 1. For example, a mixture equation that is
sometimes used is:

$$P = P_1 \phi_1^2 + P_2 \phi_2^2. \tag{2}$$

This equation is identical to

$$P = P_1 \phi_1 + P_2 \phi_2 - (P_1 + P_2) \phi_1 \phi_2 \tag{3}$$

Thus, equations 2 and 3 are the same as equation 1 for
the special case where $I = -(P_1 + P_2)$. Another example
is the Gordon-Taylor [4] equation:

$$K_1 \phi_1 (P - P_1) + K_2 \phi_2 (P - P_2) = 0 \tag{4}$$

where K_1 and K_2 are constants. This can be rearranged
to give

$$P = \frac{K_1 \phi_1 P_1 + K_2 \phi_2 P_2}{K_1 \phi_1 + K_2 \phi_2}. \tag{5}$$

If $K_1 = K_2$, then $P = P_1\phi_1 + P_2\phi_2$, which is equation 1
when the interaction term is zero. Even if K_1 does not
equal K_2, equation 4 can roughly approximate equation 1
in some cases.

Equation 2 is a special case of a more general equation

$$P = P_1\phi_1^m + P_2\phi_2^m . \tag{6}$$

In this equation, m is a constant. If $m = +1$, then equation 6 gives the usual "rule of mixtures." However, m
can vary between $-\infty$ and $+\infty$. Then this infinite number
of equations can fill the entire space between $P = 0$ and
$P = \infty$ on the vertical axis and between $\phi_2 = 0$ and $\phi_2 = 1$
on the horizontal axis of a plot of P versus ϕ_2.

One-phase mixture equations generally should be symmetrical with respect to ϕ_1 and ϕ_2. As long as the two
components are miscible with one another, it does not
matter which one is designated as material 1. A very
general and useful equation which fulfills this condition
is:

$$P^n = P_1^n\phi_1 + P_2^n\phi_2 . \tag{7}$$

This equation, which is the same as equation 10 of Chapter
1, will be discussed in more detail in Chapter 4.

III. GRAPHICAL ILLUSTRATIONS OF THE MIXTURE EQUATIONS

The effect of the interaction term I on the property
P can best be illustrated by graphs made from equation 1
for specific cases. Three such cases are shown in Figures
1-3.

In Figure 1, the value of the property is the same
for each of the two pure components. If there is no

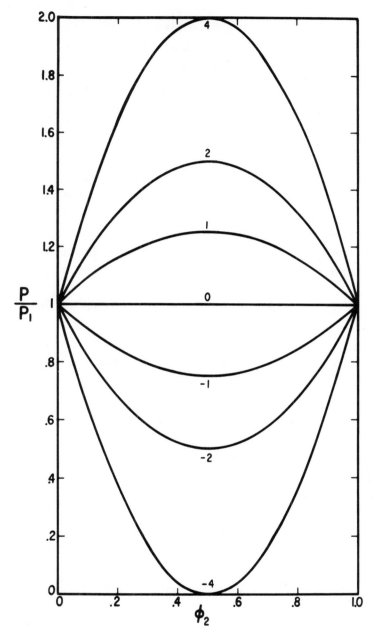

Figure 1.   Plots of equation 1 with $P_1 = P_2$. Numbers on
the curves refer to the value of I.

interaction between the components, that is, $I = 0$, a
horizontal line results in a plot of P/P_1 versus the
concentration ϕ_2. When the interaction is positive, the
curves bulge upward so that the value of the property
becomes greater than that of the components. This increase in P increases as I increases. When the interaction term is negative, the curves sag below the horizontal line, and P becomes less than either P_1 or P_2.

In Figure 2, the value of P_2 is assumed to be twice
that of P_1. Again, when $I = 0$, the "rule of mixtures"
equation holds so that a straight line connects the points
representing the pure components on a plot of P/P_1 versus
ϕ_2. Interaction produces bulges or depressions in the
curves, and the curves are no longer symmetrical about
$\phi_2 = 0.5$

In Figure 3, the value of P_2 is assumed to be ten
times that of P_1. Again, interaction produces either
positive or negative deviations from the "rule of mixtures" value. Large values of I are required to produce
a factor of two change in the value predicted by the
"rule of mixtures" equation. Thus, the greater the ratio
P_2/P_1, the larger I must be to produce a given relative
change in P/P_1.

Linear scales of P/P_1 have been used in Figures 1-3.
If a logarithmic scale is used for P/P_1, the curves of
Figure 3 become distorted as shown in Figure 4. The
aesthetic symmetry of the curves in Figure 3 disappears
in Figure 4. Note that the curves in Figures 1-3 have a
symmetry determined by a rotation of 180 degrees in the
plane of the figures about an axis perpendicular to the
plane. The curve for a given positive value of I becomes
the curve for the same negative value of I when a rotation
of 180 degrees is made.

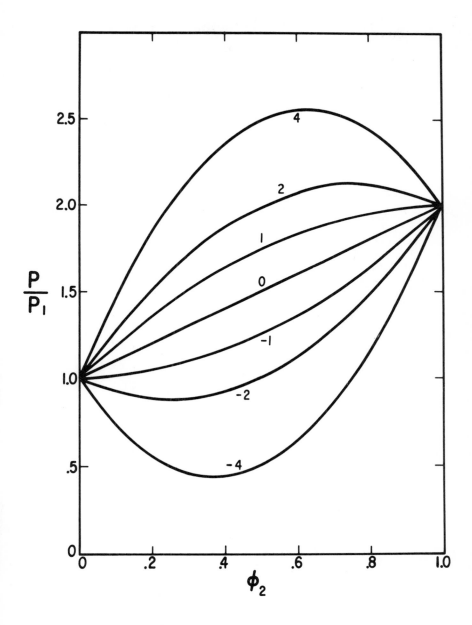

Figure 2.   Plots of equation 1 with $P_2/P_1 = 2$.   Numbers on the curves refer to the value of I.

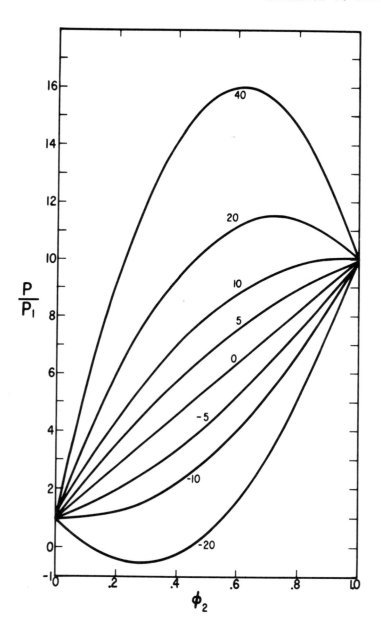

Figure 3. Plots of equation 1 with $P_2/P_1 = 10$. Numbers refer to the value of I.

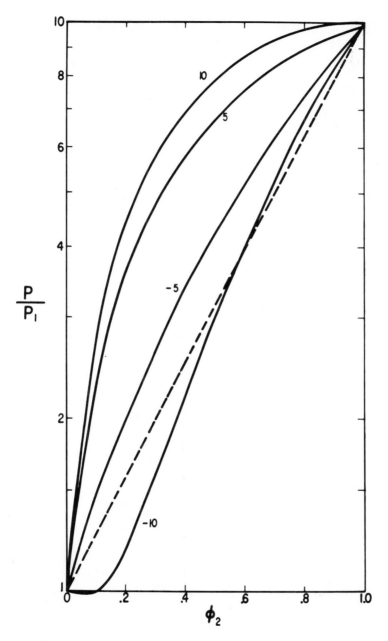

Figure 4. Equation 1 plotted with different values of I using a log P/P_1 scale for $P_2/P_1 = 10$. The dashed line is the logarithmic rule of mixtures.

Likewise, equation 6 can be plotted to give the
curves shown in Figures 5 to 7. In Figure 5, $P_1 = P_2$.
The rectangle between $P/P_1 = 0$ and $P/P_1 = 2$ can be completely filled by an infinite number of curves as m
changes from zero to infinity. As shown in Figure 6,
the space above $P/P_1 = 2$ can be filled by letting m vary
from zero to $-\infty$.

In Figure 7 it has been assumed that $P_2/P_1 = 2$. In
this case all the space in the graph between $P/P_1 = 0$
and $P/P_1 = 3$ can be filled by letting m vary from zero
to infinity. The shapes of the curves obtained from
plots of equations 1 and 6 are quite different, as can
be seen from a comparison of Figure 1 with Figure 5 and
Figure 2 with Figure 7. This difference in equations 1
and 6 is further illustrated in Figure 8 where curves
from the two equations are matched at $\phi_2 = 0$, $\phi_2 = 0.5$,
and $\phi_2 = 1$ for the case where $P_2/P_1 = 1$. The curves in
Figure 8 are given by the following equations:

A. $P = P_1 \phi_1^{0.5} + P_2 \phi_2^{0.5}$

B. $P = P_1 \phi_1 + P_2 \phi_2 + 1.655\ \phi_1 \phi_2$

C. $P = P_1 = P_2$

D. $P = P_1 \phi_1 + P_2 \phi_2 - 2\phi_1 \phi_2 \equiv P_1 \phi_1^2 + P_2 \phi_2^2$

E. $P = P_1 \phi_1 + P_2 \phi_2 - 3.5\phi_1 \phi_2$

F. $P = P_1 \phi_1^4 + P_2 \phi_2^4$

Examples of curves derived from equation 1 are B, D,
and E, while curves derived from equation 6 are A, D,
and F. Only in the case of curve D does equation 6 give
the same result as equation 1. The match between curves
A and B is less satisfactory, while the match between
curves E and F is better but not perfect.

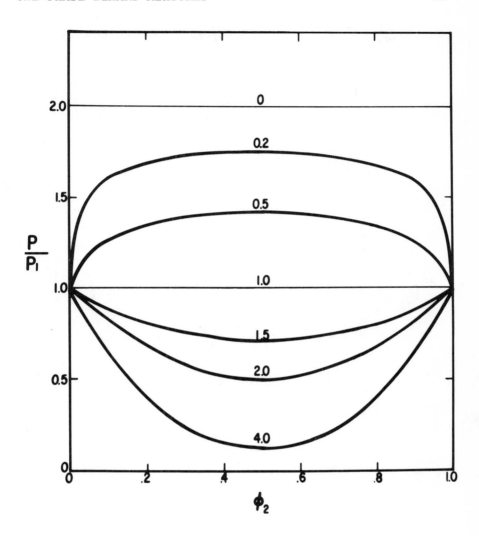

Figure 5.   Equation 6 plotted for different positive
             values of m when $P_1 = P_2$.

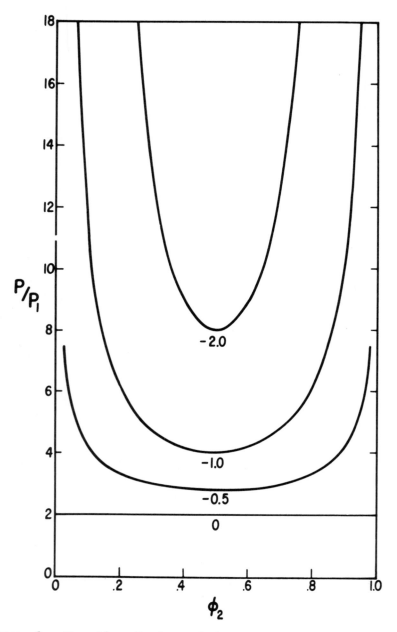

Figure 6.   Equation 6 plotted for different negative
            values of m when $P_2/P_1 = 1$.

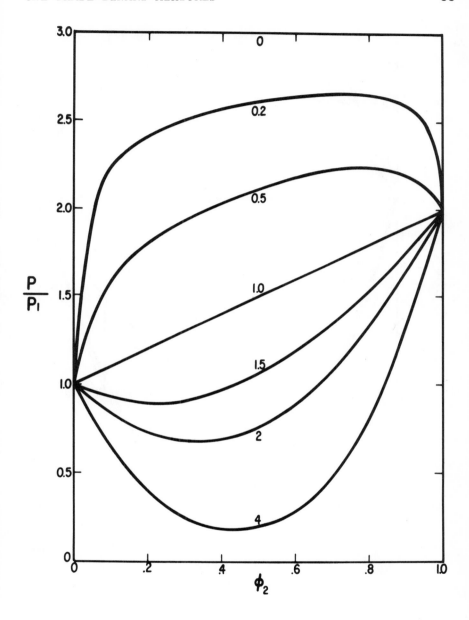

Figure 7.   Equation 6 plotted for different values of m
            when $P_2/P_1 = 2$.

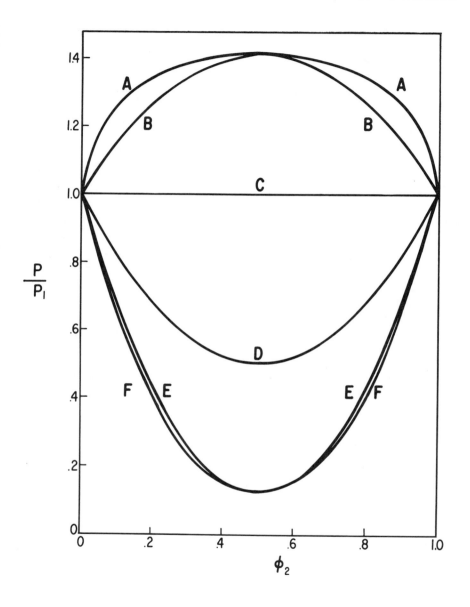

Figure 8.   A comparison of equations 1 and 6 for $P_1 = P_2$.
Curves A and B and curves E and F are matched
at $\phi_2 = 0.5$.   See text for description of
curves.

IV.   ILLUSTRATIVE PRACTICAL EXAMPLES

A few examples from the literature will be used to illustrate the prediction of properties by the use of such mixture rules as equation 1.

Figure 9 shows the viscosity of mixtures of cyclo-hexane and heptane at 37.8°C [5]. Large interactions would not be expected with this system, so the volume fraction ϕ_2 of heptane should be the proper concentration variable. (The original data were given in terms of mole fractions, but the viscosity-mole fraction plot was more curved than the plot shown in Figure 9.) In order to use equation 1, one piece of information in addition to the viscosity of the pure components is required. In the construction of Figure 9, the value of the viscosity at $\phi_2 = 0.5$ was used to derive the value of I. The resulting equation is:

$$\eta = 0.510\ \phi_1 + 0.947\ \phi_2 - 0.30\ \phi_1\phi_2. \quad (8)$$

The fit of the curve to the experimental points is quite good but not perfect.

The second example is the calculation of the critical temperature of a hydrocarbon mixture. The results in Figure 10 have been chosen to be similar to the critical temperature data for a mixture of ethane and butane. Critical temperatures have been calculated accurately by the use of an equation with distorted concentration scales similar to equation 18 of Chapter 1 [6]. The actual equation that was used is:

$$T_c = \frac{T_1}{1 + C_1\ X_2/X_1} + \frac{T_2}{1 + C_2\ X_1/X_2} \quad (9)$$

where T_c is the critical temperature of the mixture,

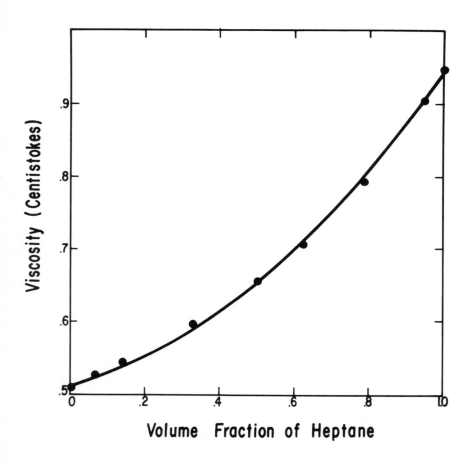

Figure 9.   Viscosity of a mixture of cyclohexane and
            heptane at 37.8°C. Comparison of equation 1
            with experimental data.

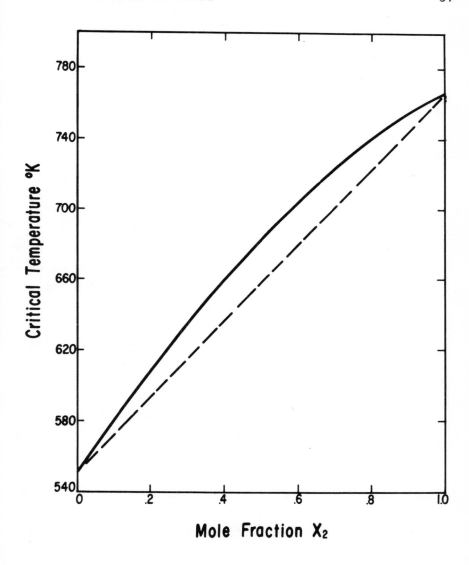

Figure 10.   The critical temperature of typical hydro-
             carbon mixtures.   Comparison of equations 1
             and 10 with equation 9.

T_1 and T_2 are the critical temperatures of the two hydro-
carbons, and C_1 and C_2 are constants. The values used
to calculate the curve in Figure 10 were:

$T_1 = 550°K$, $T_2 = 766°K$, $C_1 = 0.90$, and $C_2 = 0.95$.

The same curve in Figure 10 was calculated also within
the experimental error using the general mixing rule,
equation 1. The equation was:

$$T_C = 550 \ X_1 + 766 \ X_2 + 96 \ X_1 X_2 \ . \tag{10}$$

The general equation for mixing as illustrated by equation 10 is much simpler than equation 9 using distorted
concentration scales. Note that equation 9 is the same
as equation 5 if $C_1 = K_2/K_1$ and $C_2 = K_1/K_2$.

Figures 11 and 12 show the density and the refractive
index of mixtures of methanol and toluene [7]. The
original concentration data were given in mole fractions,
but the property-mole fraction plots gave curves which
would lead one to believe there were strong interactions
between the methanol and toluene molecules because of the
satisfactory fit of the data to equation 1. However,
plots of the data against volume fraction of methanol
gave straight lines within the experimental error. These
straight lines would indicate that the interactions between the different types of molecules are small. This
difference in the two curves illustrates a common problem
in calculating the properties of mixtures. If there are
no interactions, the proper concentration variable would
be volume fraction. If there are interactions, the proper concentration variable may be the mole fraction. One
may have to use both types of concentrations to tell
which is the better. The straight lines in Figures 11

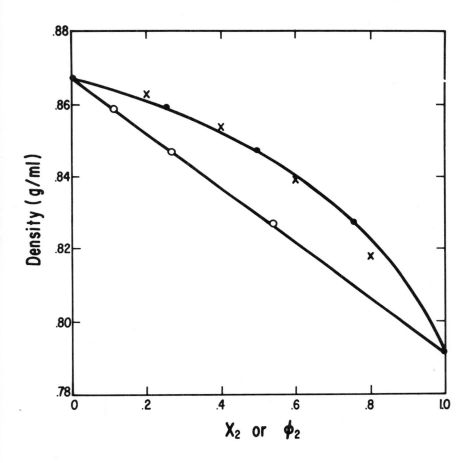

Figure 11.   Density of mixtures of methanol and toluene.
Open circles are experimental data plotted
as a function of volume fraction while the
closed circles are the same data plotted as
a function of mole fraction.   X = points
calculated using mole fractions in equation 1.

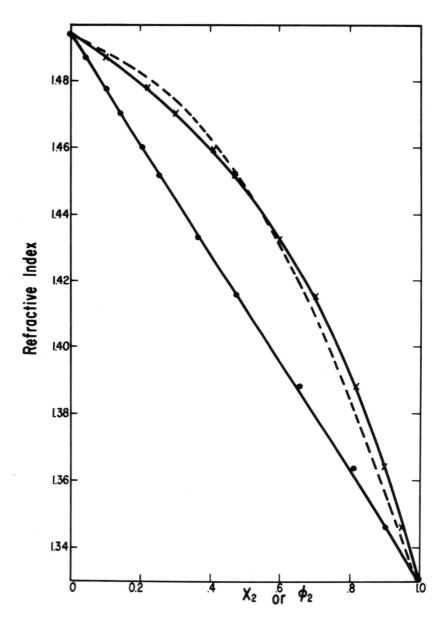

Figure 12.   Refractive index of mixtures of methanol and
    toluene.   Closed circles are the experimental data
    plotted as a function of the volume fraction of methanol
    while the crosses are the same data plotted as a function
    of mole fraction.   The dashed line is an attempt to match
    the mole fraction results with equation 1.

and 12 are the "rule of mixtures" using volume fractions. The experimental data on density ρ in Figure 11 using the mole fraction of methanol as the concentration variable also can be approximated by the equation

$$\rho = 0.8669\ X_1 + 0.7913\ X_2 + 0.072\ X_1 X_2. \quad (11)$$

The refractive index R in Figure 12, also at 25°C, can be approximated by

$$R = 1.494\ X_1 + 1.330\ X_2 + 0.140\ X_1 X_2, \quad (12)$$

however, better fit to the experimental data in both cases is obtained by using volume fraction rather than mole fraction.

The thermodynamic theory of phase diagrams is well established. However, very often not all the necessary thermodynamic information is available, so a more empirical approach becomes useful. An example of such an approach is shown in Figure 13 where the liquid-vapor phase diagram for mixtures of benzene and toluene at one atmosphere of pressure is shown [8]. At a constant temperature, the composition of the liquid and vapor phases is given by the curves marked L and V, respectively. At a given composition, the same curves give the bubble temperature T_V and the condensation temperature T_L (the lower curve). These temperatures are given approximately by

$$T_L = 110.7\ X_1 - 80.6\ X_2 - 13\ X_1 X_2 \quad (13)$$

$$T_V = 110.7\ X_1 - 80.6\ X_2 + 13\ X_1 X_2 . \quad (14)$$

This fairly ideal system might be expected to show symmetry in which the same interaction factor is used in both equations with just a change in sign.

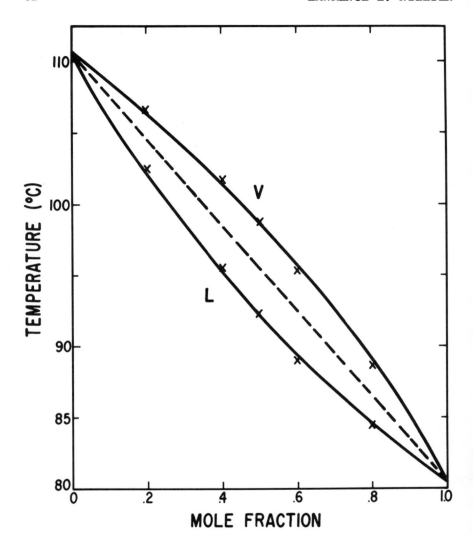

Figure 13.   The benzene-toluene boiling point diagram.
              V is the vapor curve while L is the liquid
              curve.   The full curves represent the experi-
              mental data.   X = values calculated by
              equation 1 by matching the experimental value
              at $X_2 = 0.5$.

The glass transition temperature T_g is often the most important factor in determining many of the mechanical and other properties of plastics and rubbery materials [9]. The T_g of a polymer can be decreased by adding a liquid or plasticizer to the polymer. Figure 14 shows that equation 1 often can be used to estimate the glass transition temperature of a polymer when a plasticizer is added to it [10]. The T_g of polystyrene-tricresyl phosphate mixtures is given by

$$T_g = 85.5 \ W_1 - 64.5 \ W_2 - 164 \ W_1 W_2 \qquad (15)$$

while T_g of polystyrene-naphthyl salicylate mixtures is given by

$$T_g = 85.5 \ W_1 - 29.5 \ W_2 - 124 \ W_1 W_2. \qquad (16)$$

These equations were obtained by matching the experimental curve at $W_2 = 0.5$ to estimate the interaction factor I of equation 1. One would expect that the volume fraction should have been used in Figure 14 rather than the weight fraction. However, if the densities of the two components are nearly the same, the volume fractions and the weight fractions have about the same value.

The glass transition temperatures of polymers can be changed also by copolymerizing two types of monomeric units. Figures 15 and 16 illustrate how equation 1 can be used to estimate the T_g of two different copolymer systems. In Figure 15 the T_g of vinyl acetate-vinyl chloride copolymers as a function of the weight fraction of vinyl chloride W_2 can be approximated by [11]:

$$T_g = 30.0 \ W_1 + 80.0 \ W_2 - 28 \ W_1 W_2. \qquad (17)$$

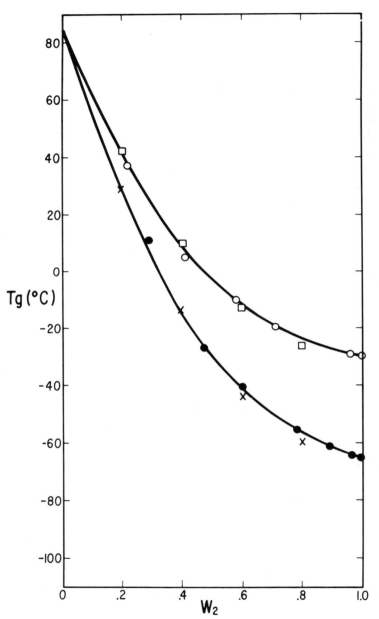

Figure 14. Glass transition temperature of plasticized
polystyrene. Tricresyl phosphate in polystyrene:
● Experimental, X Equation 1. Naphthyl salicylate in
polystyrene: O Experimental, □ Equation 1. W_2 = weight
fraction of plasticizer.

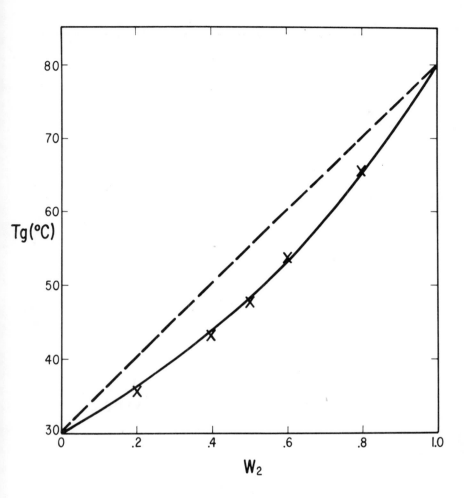

Figure 15.   Glass transition temperature of vinyl acetate-
             vinyl chloride copolymers.   Solid line repre-
             sents the experimental curve.   Values from
             equation 1 are represented by X. W_2 =
             weight fraction of vinyl chloride.

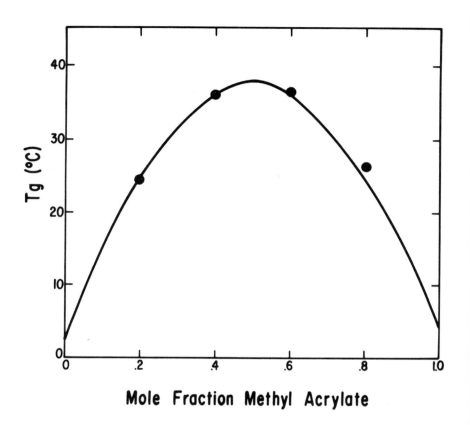

Figure 16.   Glass transition temperature of methyl
             acrylate-vinylidene chloride copolymers.
             Solid line = experimental curve. ● =
             calculated by equation 1.

In Figure 16 the T_g of vinylidene chloride-methyl acryl-
ate copolymers can be estimated from [12]:

$$T_g = 2.5 \, X_1 + 5.0 \, X_2 + 136 \, X_1 X_2 \qquad (18)$$

In Figure 16 the experimental values of T_g deviate strongly from the "rule of mixtures" equation, but the values
can be estimated reasonably well when a large interaction
term is included.

The examples used in this section illustrate the
versatility of equation 1 for calculating many properties
of mixtures which form a single phase. This general mixture equation can be used when data are available for at
least three points on the composition scale. At times,
some other mixture equation may provide better fit to the
experimental data, but in general, the equation must be
more complex than equation 1.

V.   REFERENCES

1.   J. R. Partington, Advanced Treatise on Physical
     Chemistry, Vol. 4, Longmans Green, London, 1953,
     pp. 73

2.   R. C. Reid and T. K. Sherwood, Properties of Gases
     and Liquids, McGraw-Hill, New York, 2nd Ed., 1966,
     pp. 509

3.   J. H. Hildebrand and R. L. Scott, The Solubility of
     Nonelectrolytes, 3rd Ed., Reinhold, New York, 1950
     pp. 40 and 141.

4.   M. Gordon and J. S. Taylor, J. Appl. Chem. (London),
     2, 493 (1952).

5.   R. A. McAllister, AIChE J., 6, 427 (1960).

6.   R. B. Grieves and G. Thodos, AIChE J., 8, 550 (1962).

7.   L. W. Hammond, K. S. Howard, and R. A. McAllister,
     J. Phys. Chem., 62, 637 (1958).

8.   W. L. Badger and W. L. McCabe, Elements of Chemical
     Engineering, 2nd Ed., McGraw-Hill, New York, 1936,
     pp. 324

9.   L. E. Nielsen, Mechanical Properties of Polymers
     and Composites, Vol. 1, Marcel Dekker, New York,
     1974.

10. E. Jenckel and R. Heusch, Kolloid Zeit, 130, 89 (1953).

11. K. H. Illers, Kolloid Zeit, 190, 16 (1963).

12. M. Hirooka and T. Kato, J. Polymer Sci., (Letters), 12, 31 (1974).

Chapter 3

TWO-PHASE MIXTURES WITH ONE CONTINUOUS PHASE

I. INTRODUCTION

Several kinds of systems fit into the category of two-phase mixtures with one continuous phase and one dispersed phase. These systems include most composite materials, suspensions, foams, and filled polymers, and high-impact polymers. Either phase can be the "hard" phase or the "soft" phase. Polymers filled with rigid filler particles are an example of a system in which the "hard" phase is the dispersed phase while foams and high-impact polymers are examples in which the "hard" phase is the matrix or the continuous phase.

With few exceptions, when there is a dispersed phase, it is not possible to cover the entire composition range between the two pure components. This is because the dispersed phase can not pack in such a manner as to completely fill a space. For example, uniform rigid spheres can never fill more than about 74 percent of a given volume. Metal alloys are an exception where all the space can be filled. Foams can be another example because the walls of the continuous phase can deform. For cases where all of the space can not be filled, maximum packing fraction ϕ_m is an important factor in

determining the properties of such mixtures. Thus, mixture rules for such systems should be capable of accomo-
dating the existence of this packing fraction, and the
effect of this packing can be accomplished by using a
reduced concentration scale [1,2].

Interaction between the constituents generally does
not occur in two-phase systems except at the interface.
Thus, the nature of the interface can be very important.
If the interaction is small at the interface, the adhesion between the phases usually is poor. If the two
components do interact with one another or if there is a
tendency for slight solubility of one component in the
other, the adhesion between the components generally is
strong. This degree of adhesion can have a great effect
on some properties, but its effect on other properties
may be small. Mechanical properties are especially sen-
sitive to the degree of adhesion.

Another factor which has a profound effect on the
properties of two-phase mixtures is the shape of the particles making up the dispersed phase [1,3-10]. Therefore,
spherical particles and rod-like particles do not produce the same changes in properties when comparisons are
made at equal concentrations. Likewise, the degree of
dispersion or the state of agglomeration of the particles
can have a large effect on the value of a property of a
mixture [11]. The state of orientation of the particles
also is an important factor [1,12]. For example, the
properties of a fiber-filled material when the fibers are
randomly oriented are not the same as when the fibers
are all aligned in one direction.

Examples of the types of properties which can be
calculated for two-phase mixtures have been given in
Table 2 of Chapter 1.

II. THEORY AND A GENERAL MIXTURE RULE

A tremendous number of equations, both theoretical and empirical, have been proposed for calculating the properties of two-phase mixtures. Over 100 equations have been proposed for calculating the viscosity of a suspension of spheres alone [13]. Many equations have been derived for predicting the dielectric constant of suspensions [6-9,14-17], the elastic moduli of composite materials [1,2,12,18-23], the thermal conductivity of mixtures [24-28], the permeability of gases and liquids through composite materials [10,29-32], and other properties. Fortunately, many of these equations can be shown to be specific examples of a very general equation for two-phase systems in which one of the phases is dispersed.

The general mixture rule which will be emphasized in this work is [1,2,33]:

$$\frac{P}{P_1} = \frac{1 + AB\phi_2}{1 - B\psi\phi_2} \tag{1}$$

where

$$A = k_E - 1 \tag{2}$$

$$B = \frac{P_2/P_1 - 1}{P_2/P_1 + A} \tag{3}$$

$$\psi \doteq 1 + \left(\frac{1 - \phi_m}{\phi_m^2}\right)\phi_2. \tag{4}$$

The subscripts 1 and 2 refer to the matrix or continuous phase and the dispersed phase, respectively. As pointed out in Chapter 1, the constant A depends upon the shape of the dispersed particles, their state of agglomeration, their orientation, and the nature of the interface; its

value can vary between zero and infinity. The constant
A is related to the generalized Einstein coefficient k_E,
which has been calculated from theory for many systems
[1,3,5-7,12,28]. The factor ψ enables one to use a reduced concentration scale to take into account the existence of the maximum packing fraction ϕ_m of the particles
[1,28,34]. The factor ψ can be approximated by a number
of equations, but equation 4 is the simplest. The maximum packing fraction is defined by

$$\phi_m = \frac{\text{True volume of the particles}}{\text{Actual volume occupied by the particles}} . \quad (5)$$

The maximum packing fraction can be estimated in a number
of ways. It can be obtained from the sedimentation volume [35] or from the bulk density of the compacted pow-
dered material making up the particles. The maximum
packing fraction also can be estimated from the amount
of a liquid which must be added to a powder to convert
it from a dry-appearing powder to a wet-appearing fluid
mass [36-38]. This technique is one commonly used in the
paint or surface coatings industry. Some values of ϕ_m
derived largely from theory are given in Table 1. The
reduced concentration $\psi\phi_2$ is plotted in Figure 1 as a
function of the volume fraction of dispersed phase ϕ_2
for several values of ϕ_m. The reduced volume fraction
$\psi\phi_2$ becomes 1.0 at $\phi_2 = \phi_m$ instead of at $\phi_2 = 1$. If the
particles can fill all the space, $\phi_m = 1.0$, and $\psi = 1.0$
at all concentrations.

Equations 1-4 are valid when the "hard" phase is
the dispersed phase. These equations also are valid in
the inverted case in which the "soft" phase is the dispersed phase, but one must use care because of negative
signs. For the inverted case in which the "soft" phase
is the dispersed phase, equations 1-3 can be rearranged
to give:

Table 1

Maximum Packing Fractions ϕ_m

Particles	Type of Packing	ϕ_m
Spheres	Hexagonal close packing	0.7405
"	Face centered cubic	0.7405
"	Body centered cubic	0.60
"	Simple cubic	0.5236
"	Random close packing	0.637
"	Random loose packing	0.601
Fibers	Parallel hexagonal	0.907
"	Parallel cubic	0.785
"	Parallel random	0.82
Cubes	Random	0.70
Rods	L/D = 4, random three dimensional (approx.)	0.625
"	L/D = 8, random three dimensional (approx.)	0.48
"	L/D = 16, random three dimensional (approx.)	0.30
"	L/D = 40, random three dimensional (approx.)	0.13
"	L/D = 70, random three dimensional (approx.)	0.065

$$\frac{P_1}{P} = \frac{1 + A_i \, B_i \, \phi_2}{1 - B_i \, \psi \, \phi_2} \tag{6}$$

where

$$A_i = 1/A \tag{7}$$

$$B_i = \frac{P_1/P_2 - 1}{P_1/P_2 + A_i} \tag{8}$$

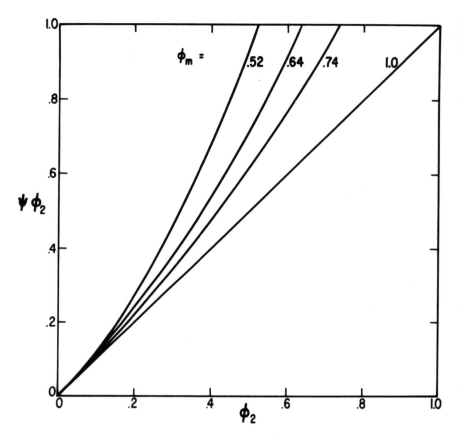

Figure 1.   Reduced concentration $\psi\phi_2$ as a function of
volume fraction of dispersed phase ϕ_2 for
different values of ϕ_m.

In equations 6-8, the subscript 1 still refers to the
continuous phase while the subscript 2 still refers to
the dispersed phase of the inverted system. Figure 2
of this chapter and Figure 3 of Chapter 1 show the high
degree of symmetry inherent in equations 1-4. These
figures are symmetrical with respect to 180° of rotation
in the plane of the paper about the center of symmetry
at $\phi_2 = 0.5$. Note that in both of these figures, logarithmic scales are used for P/P_1, and that $\psi = 1.0$.

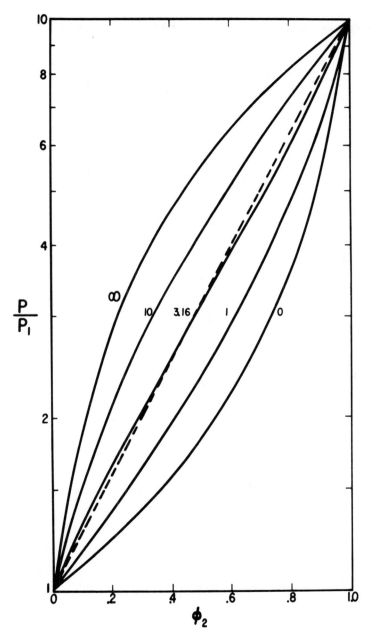

Figure 2. P/P_1 as a function of ϕ_2 for various values of
the constant A when $P_2/P_1 = 10$.   The broken
line is the logarithmic rule of mixtures.

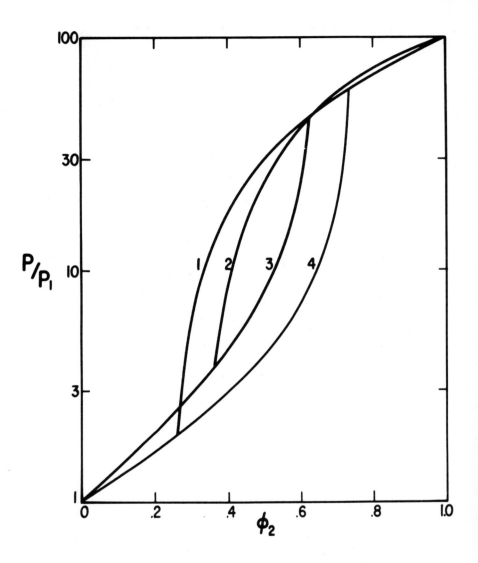

Figure 3. P/P_1 versus ϕ_2 when $P_2/P_1 = 100$.   1. A = 1.5
($A_i = 0.667$), $\phi_m = 0.74$.   2. A = 3.0 ($A_i =$
0.333), $\phi_m = 0.64$.   3. A = 3.0, $\phi_m = 0.64$.
4. A = 1.5, $\phi_m = 0.74$.

Figure 3 shows four cases (two regular and two inverted) where $\phi_m \neq 1$. When ϕ_m is less than 1.0, the curves start with one of the pure components, but they end at ϕ_m along the concentration scale before reaching the other pure component.

The values of A (or the Einstein coefficient k_E) can be obtained from theory in many cases, but in other cases the values must be determined empirically from experimental data. The original Einstein equation was derived for the viscosity of suspensions of rigid spheres [39]. In general, the equation has the form

$$\frac{P}{P_1} = 1 + k_E \ \phi_2 . \tag{9}$$

This equation, which is valid only at low concentrations, is easily shown to be the limiting form of equation 1 at very low concentrations when P_2/P_1 is very large. Thus, k_E can be obtained from the slope of the P/P_1 versus ϕ_2 plot as ϕ_2 approaches zero if P_2/P_1 is large. This is a simple empirical method of determining k_E from experimental data.

The Einstein coefficient is not the same for mechanical properties as it is for electrical properties and for thermal conductivity of mixtures. For example, for the viscosity of suspensions or for the elastic modulus of composite materials, k_E is 2.5 (A = 1.5) for dispersed spheres. For the dielectric constant, the electrical conductivity, and the thermal conductivity of mixtures containing dispersed spheres, k_E is 3.0 or A = 2.0. Many other values of k_E have been calculated for mechanical [1,4,5,12] and for electrical [6-9] properties of mixtures containing particles of different shapes and orientations. In general the values of k_E are slightly greater for electrical and thermal properties than for mechanical

properties. However, large errors will not result if the
values k_E for mechanical properties are used in place of
the values of k_E for the electrical or thermal values
most cases. Some values of k_E for mechanical properties
are given in Table 2. The corresponding values of A are

Table 2

Einstein Coefficients, k_E (Mechanical)

Type of Dispersed Phase	Orientation of Particles and Type of Stress	k_E
Dispersed spheres	Any. No slippage	2.50
Dispersed spheres	Any. Slippage	1.0
Spherical aggregates of spheres	Any. $\phi_a = \phi_m$ of spheres within aggregate	$2.5/\phi_a$
Cubes	Random (Approximate)	3.1
Uniaxially oriented fibers	Fibers parallel to tensile stress component	2L/D
Uniaxially oriented fibers	Fibers perpendicular to tensile stress component	1.50
Uniaxially oriented fibers	Longitudinal-transverse shear	2.0
Uniaxially oriented fibers	Transverse-transverse shear	1.5
Uniaxially oriented fibers	Bulk modulus	1.0
Fibers randomly oriented in three dimension	Shear. (Approximate)	L/2D

easily obtained for any of these cases by using equation
2.

The values for properties of a mixture increase with
increase in both A and P_2/P_1 when the other factors are
held constant. Figure 4 shows how P/P_1 increases as a

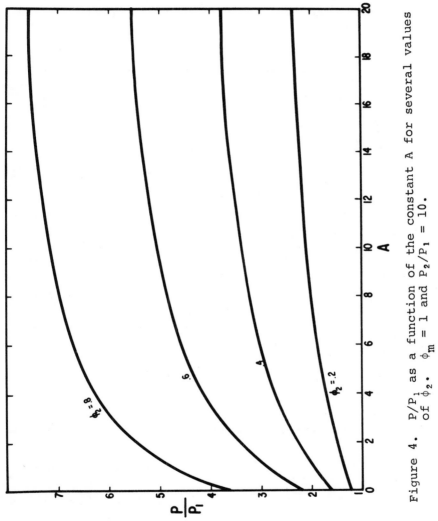

Figure 4. P/P_1 as a function of the constant A for several values of ϕ_2. $\phi_m = 1$ and $P_2/P_1 = 10$.

function of A for several values of concentration ϕ_2, assuming that $\phi_m = 1$ and $P_2/P_1 = 10$. All the curves level out to become constant at large values of A. At low values of A, the rate of increase of P/P_1 is much greater when ϕ_2 is large than when ϕ_2 is small. Figure 5 illustrates the behavior of P/P_1 as a function of P_2/P_1 for two concentrations. Again, P/P_1 reaches an asymptotic value at large values of the ratio P_2/P_1. The effect of P_2/P_1 is most pronounced at high values of the concentration. In Figure 5, it was assumed that A = 1.5 and $\phi_m = 1.0$.

Equation 1 can be put into a number of alternate forms. One of these equations is the Kerner equation [18]. This equation, which assumes $\phi_m = 1$, is:

$$\frac{P}{P_1} = \frac{\dfrac{P_2 \phi_2}{K_1 P_1 + K_2 P_2} + \dfrac{\phi_1}{K_3}}{\dfrac{P_1 \phi_2}{K_1 P_1 + K_2 P_2} + \dfrac{\phi_1}{K_3}}. \tag{10}$$

K_1, K_2, and K_3 are constants. For spheres, $A = K_1/K_2$. Another form of the Kerner equation is the Takayanagi [40] equation:

$$\frac{P}{P_1} = \frac{K_1 P_1 + K_2 P_2 - K_1 (P_1 - P_2) \phi_2}{K_1 P_1 + K_2 P_2 + K_2 (P_1 - P_2) \phi_2}. \tag{11}$$

The Kerner equation is equivalent to the Hashin-Shtrikman equation [20,41]:

$$P = P_1 + \frac{\phi_2}{\dfrac{1}{P_2 - P_1} + \dfrac{\phi_1}{k_E P_1}}. \tag{12}$$

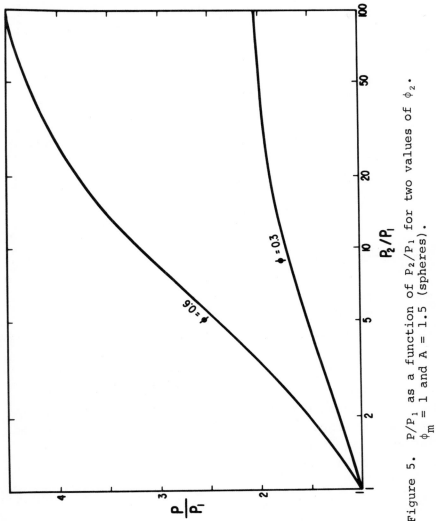

Figure 5. P/P$_1$ as a function of P$_2$/P$_1$ for two values of ϕ_2.
ϕ_m = 1 and A = 1.5 (spheres).

Another form of equation 1 in special cases is the Gordon-Taylor [42] equation:

$$k_1 \ \phi_1 (P-P_1) + k_2 \ \phi_2 (P-P_2) = 0. \qquad (13)$$

This can be seen by letting $k = k_2/k_1$ and rearranging to give

$$\frac{P}{P_1} = \frac{1 + (k \ P_2/P_1 - 1)\phi_2}{1 - (1-k)\phi_2} . \qquad (14)$$

Hoftyzer and van Krevelen [43] have shown that equation 14 can be put in the following form:

$$\frac{P_1 - P}{P_1 - P_2} = \frac{\phi_2}{1 - \phi_1(1 - 1/k)} . \qquad (15)$$

Although equations 10-15 may appear quite different from one another, they all are essentially equivalent in some respects to equation 1 with $\psi = 1$, that is, with $\phi_m = 1$. Equation 13 was proposed originally for calculating the glass transition temperature of one phase systems. In this case, k of equation 13 is related to the interaction term of equation 1 of Chapter 2. Also, because of the symmetry in the subscripts 1 and 2, equation 13 might be expected to be valid in some cases where there are two continuous phases as well. Thus, depending upon the theoretical significance of the constants k_1 and k_2, equation 13 is a very general equation.

III. ILLUSTRATIVE PRACTICAL EXAMPLES

During the last few years, the mechanical properties of composite materials have been investigated intensively. Many equations have been derived for the elastic moduli of various kinds of composite materials. However, equation 1 is the most versatile of these equations, and in most cases the predictions of equation 1 agree well with experimental results. Equation 1 also has an advantage over most other equations; if the theoretical values for A and ϕ_m are not available, these constants can be determined easily by direct experimentation or by empirical fitting of the data to curves.

Figure 6 shows the relative shear modulus G/G_1 of glass beads (75μ-90μ) in an epoxy resin [2]. The value of A = 1.5 is what would be expected for spheres, but the value of ϕ_m = 0.55, which fits the data well, is somewhat less than the expected value for spheres. The experimental value of G_2/G_1 is about 25.

Two examples for the thermal conductivity of composite materials are shown in Figure 7 [27]. Curve A is for MgO (magnesium oxide) in polystyrene while curve B is for MgO in polyethylene. The use of A = 3 and ϕ_m = 0.64 in equations 1-4 produces a curve which fits the experimental values very well. A value of A = 3 is reasonable because MgO particles are irregularly shaped and have a rough surface. Many compacted powders have maximum packing fractions near 0.64. Thus, although the original workers did not report A and ϕ_m along with their values for the thermal conductivity of these filled polymers, the experimental values can be fitted using reasonable values for A and ϕ_m.

Figure 8 shows the dielectric constant of iron particles in mineral oil [16,17]. The solid curve derived from equations 1-4 fits the experimental points very well. The broken line is the Bruggeman [14] equation which is often considered one of the better equations for predicting the dielectric constant ε of mixtures. The Bruggeman equation, which is

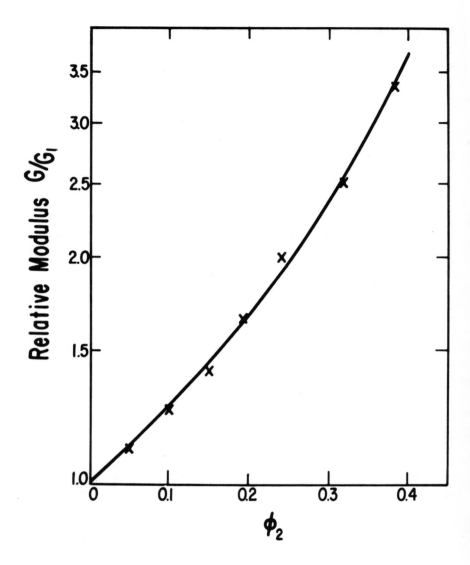

Figure 6. Relative shear modulus as a function of
           volume fraction of glass beads in an epoxy
           resin. Curve is equation 1 with A = 1.5,
           ϕ_m = 0.55, and G_2/G_1 = 25. X = experimental
           values using beads with a diameter of 75 to 90
           micrometers. [Data from Reference 2.]

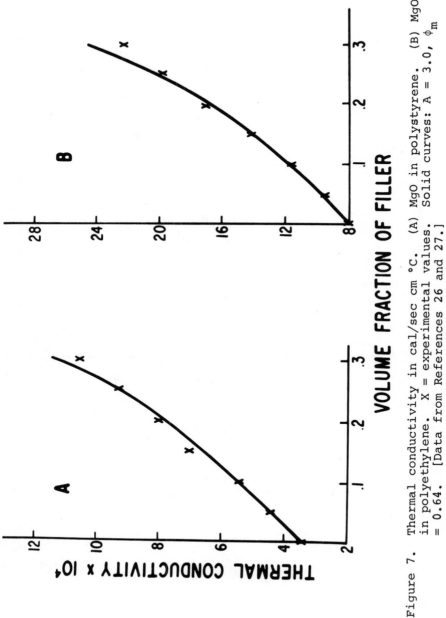

Figure 7. Thermal conductivity in cal/sec cm °C.   (A) MgO in polystyrene.   (B) MgO in polyethylene. X = experimental values. Solid curves: A = 3.0, ϕ_m = 0.64. [Data from References 26 and 27.]

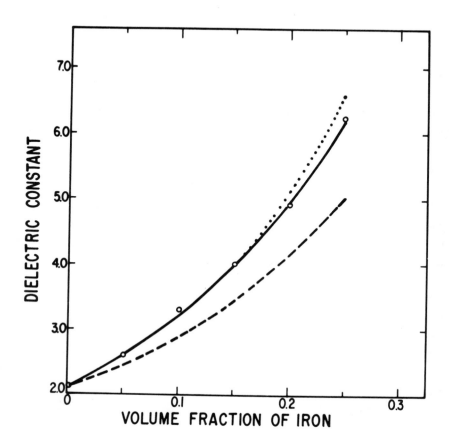

Figure 8.   Dielectric constant of iron particles dis-
            persed in mineral oil. Solid line is
            equation 1 with A = 3.5 and ϕ_m = 0.52.
            Broken line: Bruggeman equation. Dotted
            line: equation of Morabin, et al. Open
            circles: Data from Reference 16. [Reprinted
            from Reference 17.]

$$\frac{\varepsilon}{\varepsilon_1} = \frac{1}{(1 - \phi_2)^3} \quad , \tag{16}$$

does not take into account either particle shape or

packing. Obviously, this equation gives a very poor prediction of the experimental values in this case. The
dotted line is based upon a theory by Morabin, et al. [15]
which does consider particle shape in an empirical manner.
This theory fits the data nearly as well as equation 1.
The solid curve based on equations 1-4 used the values
A = 3.5 and ϕ_m = 0.52. This value of A indicates that
the iron particles were short rods of length several
times longer than the diameter. The value of ϕ_m = 0.52
also is consistant with short rods, which do not pack as
tightly as particles which are more nearly spherical in
shape. The data in both Figures 7 and 8 had to be fitted
empirically to the mixture equation by the author, which
illustrates a fault that is common to nearly all the data
found in the literature. Not all the pertinent data are
included in the original work. For example, values of
particle shape, of A (or k_E), and of ϕ_m are seldom given
even though this information is essential to properly
test such equations as those used in this chapter.

Two examples of inverted systems will be given which
make use of equations 6-8. The first of these is the
shear modulus of rigid polystyrene containing a dispersed
rubber phase to give a high-impact strength material.
Cigna [44] studied many of these polyblend systems, which
contain a soft material dispersed in a hard one, and
empirically drew a curve through the experimental points
of a plot of the shear modulus versus volume fraction of
rubber. This curve is shown in Figure 9. Up to about
30 volume percent rubber, electron microscope studies
indicated that the rubber was mostly in the shape of
spheres. The curve in Figure 9 can be fitted very accurately by using A_i = 0.86 and ϕ_m = 0.55 in equations 4 and
6-8. A value of A_i = 0.86 is consistent for spheres in
a matrix with a Poisson's ratio of 0.35 for polystyrene
[45]. (A_i would be 1/1.5 if Poisson's ratio were 0.5.)

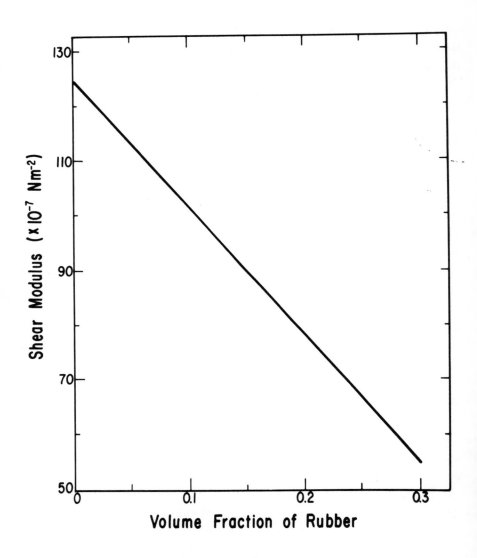

Figure 9.   Shear modulus of high-impact polystyrene
             containing rubber as a dispersed phase.
             The curve represents both the curve found
             by Cigna (Reference 44) and equation 1
             with A_i = 0.86 (spheres), Poisson's ratio
             of 0.35 for polystyrene, and ϕ_m = 0.55

The value of $\phi_m = 0.55$ might indicate that the spheres of rubber were partly agglomerated into clusters, as is often the case with particles which have a tendency to stick together.

The second example of an inverted system is the permeability of plastics to gases or liquids when filled with platelike particles which are impermeable. This system behaves as an inverted system because the permeability decreases as the concentration of the dispersed phase increases. This is in contrast to the usual case, such as the elastic modulus, where the value of the property of the mixture increases with the concentration of the dispersed phase. In this case of permeability, $P_2/P_1 = 0$ where P_1 is the permeability of the matrix material to the gas or liquid. Nielsen [32] derived the following equation for such systems:

$$\frac{P_1}{P} \doteq \frac{1 + (L/2t)\phi_2}{\phi_1} . \tag{17}$$

In this equation L is the width of a face of the filler particle, and t is the thickness of the filler plates. Equation 17 is of the same form as equation 6 with $A_i = L/2t$ since $\phi_1 = 1 - \phi_2$ and $B_i = 1.0$. Platelike particles oriented in a common plane can pack to fill nearly all of a space, so $\phi_m \doteq 1$. Thus, platelike fillers can greatly reduce the permeability of a plastic. The decrease in permeability is greatest for fillers with a large aspect ratio L/t. When the aspect ratio is 1.0, the particles are cubes. For cubes, equation 17 becomes identical with the Maxwell (29) equation for the permeability of a composite material containing spheres. The Maxwell equation is generally not put in the form of equation 17 but is given as

$$\frac{P}{P_1} = \frac{2(1 - \phi_2)}{3 - (1 - \phi_2)} \cdot \tag{18}$$

Equations 1-4 or 6-8 have been used to test hundreds of systems with a dispersed phase in a matrix material. Generally, these equations can be made to fit the experimental data quite well using reasonable values for A and ϕ_m if values for these constants were not independently available.

IV. REFERENCES

1. L. E. Nielsen, Mechanical Properties of Polymers and Composites, Vol. 2, Marcel Dekker, New York, 1974.

2. T. B. Lewis and L. E. Nielsen, J. Appl. Polymer Sci., 14, 1449 (1970).

3. J. Happel and H. Brenner, Low Reynolds Number Hydrodynamics, Prentice-Hall, New York, 1965.

4. J. M. Burgers, Second Report on Viscosity and Plasticity, North Holland, Amsterdam, 1938.

5. H. L. Frisch and R. Simha, Rheology, Vol. 1, F. R. Eirich, Ed., Academic Press, New York, 1956.

6. L. K. H. van Beek, Progress in Dielectrics, Vol. 7, J. Birks, Ed., Iliffe, London, 1967.

7. C. T. O'Konski, J. Phys. Chem., 64, 605 (1960).

8. R. W. Sillars, J. IEE, 80, 378 (1937).

9. A. R. Von Hippel, Dielectrics and Waves, Wiley, New York, 1954.

10. B. S. Mehta, A. T. DiBenedetto, and J. L. Kardos, Internat. J. Polym. Mater., 3, 269 (1975).

11. T. B. Lewis and L. E. Nielsen, Trans. Soc. Rheol., 12, 421 (1968).

12. J. E. Ashton, J. C. Halpin, and P. H. Petit, Primer on Composite Analysis, Technomic, Stamford, Conn., 1969.

13. I. R. Rutgers, Rheol. Acta, 2, 202, 305 (1962).

14. D. A. G. Bruggeman, Ann. Phys. Lpz., 24, 636 (1935).

15. A. Morabin, A. Tete, and R. Santini, Rev. Gen. Elect., 76, 1054 (1967).

16. K. Lal, R. Parshad, J. Phys., D6, 1788 (1973).

17. L. E. Nielsen, J. Phys., D7, 1549 (1974).

18. E. H. Kerner, Proc. Phys. Soc., 69B, 808 (1956).

19. Z. Hashin, Appl. Mech. Rev., 17, 1 (1964).

20. Z. Hashin and S. Shtrikman, J. Mech. Phys. Solids, 11, 127 (1963).

21. C. van der Poel, Rheol. Acta, 1, 198 (1958).

22. B. Budiansky, J. Mech. Phys. Solids, 13, 223 (1965).

23. J. C. Smith, J. Res. Nat. Bur. Stand., 79A, 419 (1975).

24. E. H. Kerner, Proc. Phys. Soc., 69B, 802 (1956).

25. S. C. Cheng and R. I. Vachon, Int. J. Heat Mass Transfer, 13, 537 (1970).

26. D. Sundstrom and Y-D. Lee, J. Appl. Polymer Sci., 16, 3159 (1972).

27. L. E. Nielsen, J. Appl. Polymer Sci., 17, 3819 (1973).

28. L. E. Nielsen, Ind. Eng. Chem. Fundamentals, 13, 17 (1974).

29. J. C. Maxwell, Electricity and Magnetism, 3rd Ed., Vol. 1, Dover, New York, 1891, pp. 440.

30. C. H. Klute, J. Appl. Polymer Sci., 1, 340 (1959).

31. R. E. Meredith and C. W. Tobias, J. Appl. Phys., 31, 1270 (1960).

72 LAWRENCE E. NIELSEN

32. L. E. Nielsen, J. Macromol. Sci., Al, 929 (1967).

33. L. E. Nielsen, J. Appl. Phys., 41, 4626 (1970).

34. J. V. Milewski, Composites, 4, 258 (1973).

35. J. V. Robinson, Trans. Soc. Rheol., 1, 15 (1957).

36. G. D. Parfitt, Dispersion of Powders In Liquids, Elsevier, Amsterdam, 1969, pp. 297.

37. T. C. Patton, Paint Flow and Pigment Dispersion, Interscience, New York, 1964, pp. 184.

38. A. I. Medalia, J. Colloid Interf. Sci., 32, 115 (1970).

39. A. Einstein, Ann. Physik, 17, 549 (1905); 19, 289 (1906); 34, 591 (1911).

40. M. Takayanogi, J. Appl. Polymer Sci., 10, 113 (1966).

41. Z. Hashin and S. Shtrikman, J. Appl. Phys., 33, 3125 (1962).

42. M. Gordon and J. S. Taylor, J. Appl. Chem. (London), 2, 493 (1952).

43. P. J. Hoftyzer and D. W. van Krevelen, Angew. Makromol. Chem., 56, 1 (1976).

44. G. Cigna, J. Appl. Polymer Sci., 14, 1781 (1970).

45. L. E. Nielsen, J. Composite Mater., 2, 120 (1968).

Chapter 4

TWO-PHASE MIXTURES WITH TWO CONTINUOUS PHASES

I. INTRODUCTION

Systems with two continuous phases may not be as
common as the types of mixtures discussed in Chapters 2
and 3, but there are more examples than one might assume.
Systems which have two continuous phases are open-celled
foams, felts or mats impregnated with gases, liquids, or
solids, many polyblends and block polymers, crystalline
polymers, laminates made up of sheets of some materials
glued together, various other kinds of interpenetrating
networks, and systems which are in the process of going
through a phase inversion. Factors, in addition to the
concentration and the properties of the constituents,
which are important in determining the properties of such
systems are the nature of the interface and the morphology
or the shape and orientation of the two phases. In many
cases the morphology is so complex that it is impossible
to describe the system in a simple quantitative manner.
Fortunately, in many cases, it appears that morphology
may play a less important role than one might be led to
believe from experience with systems which contain only
one continuous phase.

Although the same mixture rules may be made to apply
to systems with either one or two continuous phases in

some cases, in general the equations for the two kinds
of two-phase mixtures must differ. The main distinguish-
ing feature between the two cases is the nature of the
symmetry of the equations. With mixtures containing only
one continuous phase, it makes a great deal of difference
which phase is the continuous one and which is the dispersed one. Thus, mixture rules for such systems generally must indicate clearly which of the phases is continuous. On the other hand, in mixture rules for systems
with two continuous phases, either phase can be designated phase 1 or phase 2. The mixture rules for such
mixtures generally are symmetrical with respect to
material 1 and material 2.

The same mixture rules which apply to systems with
two continuous phases may be applicable also to miscible
mixtures which have a single phase. The same kind of
symmetry requirements apply to one phase systems as to
mixtures with two continuous phases because it makes no
difference which material is designated as material 1
and which is material 2. Thus, much of the discussion
in this chapter can be applied to single phase systems
also.

A phase need not necessarily be continuous to behave
as though it were continuous. An example is long fibers
aligned parallel to the direction of an applied force in
a composite system. If the fibers are several hundred
times as long as their diameters, they behave as though
they are continuous even though their length may be only
a small fraction of the size of the test specimen. Likewise, a mat made up of long fibers behaves as though the
mat were a continuous phase even though the length of the
individual fibers is small compared to the dimensions of
the mat. The fibers do contact one another, and thus
they form a continuous structure. Such a structure is
said to have "connectivity."

II. MIXTURE RULES

There are a number of equations which have the proper symmetry and which are capable of filling the rectangular space of a plot of a property versus composition with an infinite number of curves. This rectangular space consists of the distance designated by $P_2 - P_1$ on the vertical axis and the distance between $\phi_2 = 0$ and $\phi_2 = 1$ on the horizontal axis. For the case of systems with two continuous phases, the mixture rule which will be emphasized is

$$P^n = P_1^n \, \phi_1 + P_2^n \, \phi_2 \ . \qquad (1)$$

In order to fill the entire space of the rectangle discussed above, the constant n must vary between $-\infty$ and $+\infty$. However, in most practical cases, the upper and lower bounds which a given property can take on are defined by values of n between $n = -1$ and $n = +1$, which values give the "inverse rule of mixtures" and the "rule of mixtures," respectively. (See Chapter 1.)

Figures 1-3 illustrate how plots of Equation 1 look for values of n between -1 and +1 for $P_2/P_1 = 2$, $P_2/P_1 = 10$, and $P_2/P_1 = 1000$. Note that the properties are plotted on a logarithmic scale. When P_2/P_1 is small, the area between the upper and lower bounds as designated by $n = +1$ and $n = -1$, respectively, is small. However, when P_2/P_1 is large, the separation between the upper and lower bounds is large also.

It can be shown after some mathematical manipulation that a special case of Equation 1 is the logarithmic rule of mixtures which results when $n = 0$:

$$\log P = \phi_1 \log P_1 + \phi_2 \log P_2. \qquad (2)$$

This equation is useful in predicting the elastic modulus of mixtures of polymers in the composition range where

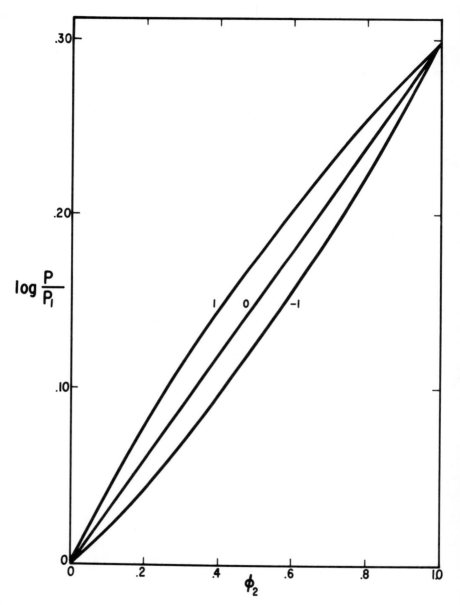

Figure 1.   Log P/P_1 as a function of ϕ_2 for various
            values of n in Equation 1. $P_2/P_1 = 2$.

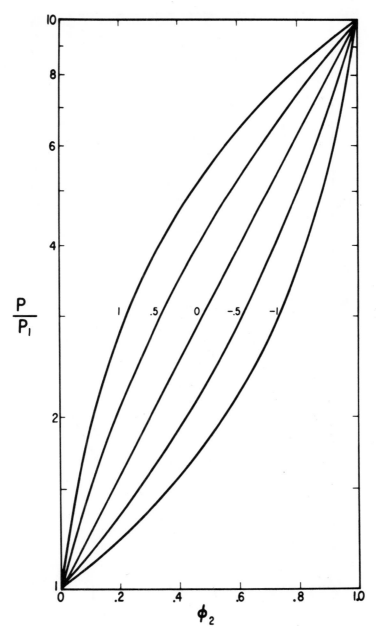

Figure 2.   P/P$_1$, plotted on a logarithmic scale, as a
   function of ϕ_2 for various values of n in Equation 1.
   P$_2$/P$_1$ = 10.

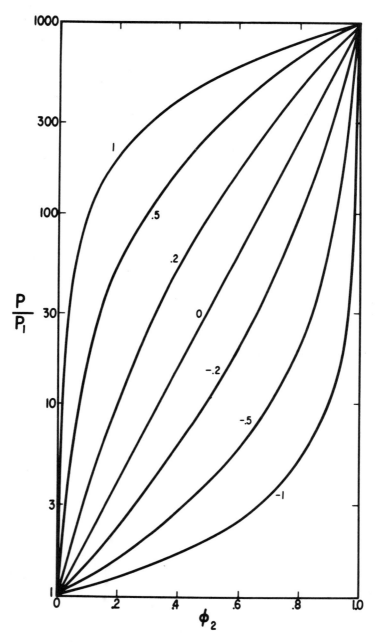

Figure 3. P/P_1 as a function of ϕ_2 for various values of n in Equation 1. $P_2/P_1 = 1000^2$. P/P_1 is plotted on a logarithmic scale.

phase inversion can occur. Equations 1 and 2 also often
can be used to estimate such properties as the viscosity
of mixtures [1]. Figures such as Figures 1-3 are symmetrical to rotations of 180 degrees in the plane of the
paper. A curve for n = +k becomes the curve for n = -k
when it is rotated 180 degrees in the plane of the graph.
This symmetry is reflected in the following equation:

$$P_{\phi_2}^{-n} = \frac{P_2/P_1}{P_{\phi_1}^{+n}} \quad . \tag{3}$$

$P_{\phi_2}^{-n}$ is the value of the property P using a negative value
of n and a concentration of ϕ_2 in Equation 1. $P_{\phi_1}^{+n}$ is the
value of the property using a positive value of n and a
concentration of $\phi_1 = 1 - \phi_2$ in Equation 1.

Although the results generally have no physical significance, the values of n can be extended beyond the
range between -1 and +1 in Equation 1. This extension
to larger values of n is illustrated in Figure 4. It is
apparent that the whole area between the coordinates of
the figure could be filled by the infinite number of
curves which could be plotted using values of n between
positive and negative infinity.

Figure 5 shows how P/P_1 varies as a function of n
for several values of ϕ_2. The value of a property increases as n and ϕ_2 increase. Figure 5 should be compared with Figure 4 of Chapter 3 to see how properties
vary as a function of the important parameters for systems with one continuous phase and for systems with two
continuous phases. The curves in Figure 5 and those in
Figure 4 of Chapter 3 have different shapes.

Many other equations are capable of filling the
space defined between the limits P_1 and P_2 and between
$\phi_2 = 0$ and $\phi_2 = 1$ in addition to Equation 1. Only two of
these equations will be mentioned. The first equation is

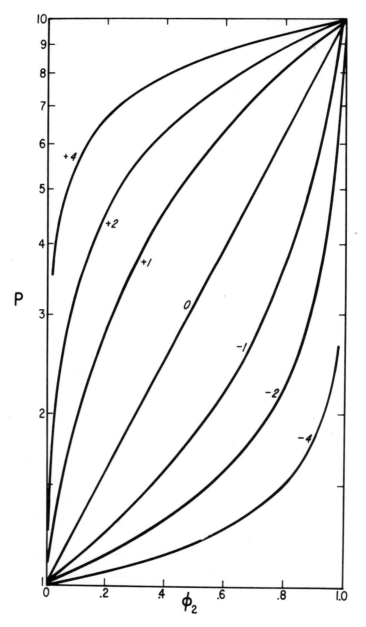

Figure 4.   Property P as a function of ϕ_2 for various
values of n in Equation 1. $P_2/P_1 = 10$.   Some values
of n are outside the limits of the normal upper and
lower bounds.

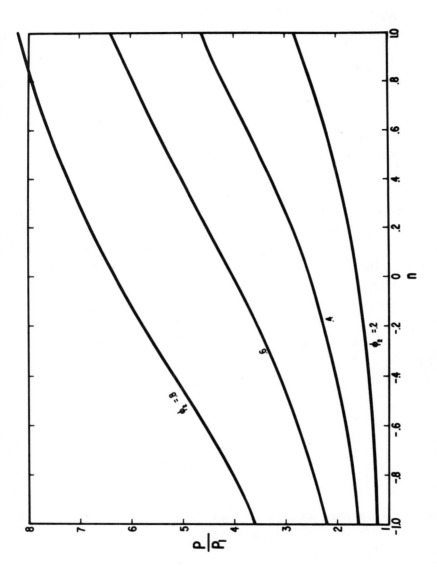

Figure 5. P/P_1 versus n for several values of ϕ_2 in Equation 1. $P_2/P_1 = 10$.

$$\frac{P}{P_1} = \left(\frac{P_2 - P_1}{P_1}\right) \phi_2^m + 1, \qquad (4)$$

which is equivalent to

$$\frac{P - P_1}{P_2 - P_1} = \phi_2^m . \qquad (5)$$

In equations 4 and 5, m is a constant which can vary between 0 and $+\infty$. Figure 6 illustrates the shape of the curves obtained from Equation 4. These curves have a different type of symmetry than is obtained using Equation 1. The curves in Figure 6 have symmetry to a rotation about the curve for m = 1 as the axis of rotation, that is, the rotation is out of the plane of the paper. If the curve for m = 1/k is rotated 180 degrees about the curve for m = 1, the curve for m = k is obtained. Note that P/P_1 is plotted on a linear scale.

The second equation is one proposed by Coran and Patel [2]:

$$\frac{P}{P_1} = \left(\frac{P_2 - P_1}{P_1}\right) f + 1 \qquad (6)$$

where

$$f = (n + 1)\phi_2^n - n\phi_2^{n+1} . \qquad (7)$$

The value of the constant n can vary between zero and infinity. Figure 7 illustrates the type of curves obtained from Equations 6 and 7. These equations have been used to fit modulus versus concentration data empirically for many types of composite materials [2].

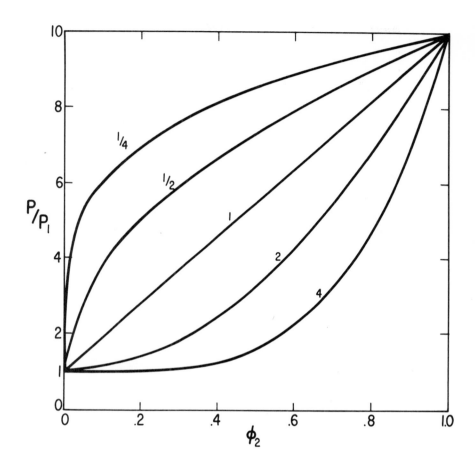

Figure 6. P/P_1 as a function of ϕ_2 for several values of m in Equation 4. $P_2/P_1 = 10$.

III. ILLUSTRATIVE PRACTICAL EXAMPLES

Crystalline polymers consist of an amorphous phase and a crystalline phase. Morphological studies indicate and mechanical studies verify that such polymers behave as though both phases are continuous. Davies' [3-5] theoretical work indicates that the elastic modulus G of

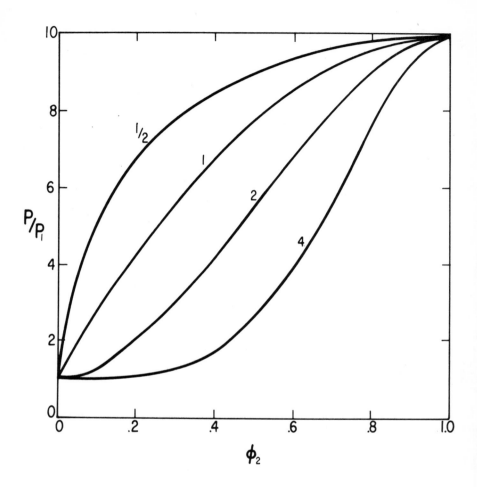

Figure 7. P/P_1 as a function of ϕ_2 for various values
         of n in Equations 6 and 7. $P_2/P_1 = 10$.

such materials should obey Equation 1 with a value of n
equal to 1/5.   That is,

$$G^{1/5} = \phi_1 \, G_1^{1/5} + \phi_2 \, G_2^{1/5} \, . \tag{8}$$

Above the glass transition temperature, the amorphous
phase behaves as a rubber with a shear modulus of about
$2.0 \times 10^5 \ Nm^{-2}$. The modulus of the crystalline phase
does not differ greatly from polymer to polymer, and it
averages out to be about $2 \times 10^9 \ Nm^{-2}$. Figure 8 shows
that Equation 8 fits the modulus-crystallinity data for
many different crystalline polymers quite well even
though the morphology of the polymers may be quite different [6]. Polymers shown in Figure 8 include poly-
ethylenes, ethylene copolymers, polypropylenes, nylons,
and many other types of semicrystalline polymers. In
Figure 8, the degree of crystallinity was measured as a
weight fraction, although volume fraction would be expected to be a better choice. Data such as those pre-
sented in Figure 8 can not be represented by the equations of Chapters 2 and 3 using reasonable values of the
parameters. The scatter in Figure 8 may be due to differences in morphology and crystallite shape and to var-
iations in G_1 and G_2 from polymer to polymer. It is re-
markable, however, that Equation 8 holds so well for
changes in modulus which differ by a factor of a thousand.

The Davies equation has been found to hold also for
interstitial polymers in which the two polymers behave
as though there were two continuous phases [5]. The
value of $n = 1/5$ was derived for the case of the elastic
modulus. However, for the dielectric constant of some
types of mixtures, Davies found that the proper value of
n is $1/3$ [3].

Mixtures of two polymers or block polymers can go
through a variety of morphologies as the concentrations
of the components change. At low concentrations of
material 2 in material 1, phase 2 tends to be dispersed
as spheres. In the neighborhood of 25 volume percent of
material 2, phase inversion can start, and both phases

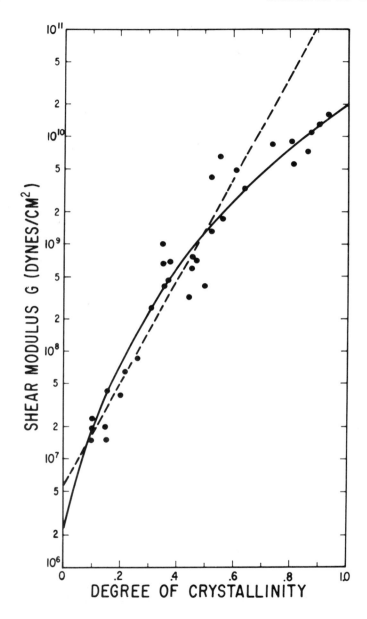

Figure 8. The shear modulus G as a function of the
  degree of crystallinity of semicrystalline polymers.
  Solid line is Equation 1; broken line is Equation 2.
  Points are experimental values for various polymers.
  [Reprinted from Reference 6.]

tend to be continuous. At high concentrations, above
about 75 percent of material 2, material 2 becomes the
continuous phase while material 1 becomes the dispersed
phase. At the very high and very low concentration
ranges, properties such as the modulus can be calculated
by the equations for dispersed systems discussed in
Chapter 3. In the intermediate region where there are
two continuous phases, the logarithmic rule of mixtures
often is very accurate [7,8]. Typical variation in modulus as a function of composition is shown in Figure 9
for the modulus of a mixture of a rigid polymer and a
rubber. A reduced concentration scale was used in the
logarithmic rule of mixtures equation in such cases.
The reduced scale starts where phase inversion starts
and ends where phase inversion is complete. The reduced
volume fractions that go into Equation 2 are:

$$\phi_1 = \frac{\phi_u - \phi}{\phi_u - \phi_L} \tag{9}$$

and

$$\phi_2 = \frac{\phi - \phi_L}{\phi_u - \phi_L} \tag{10}$$

where ϕ_L is the volume fraction where phase inversion
starts, and ϕ_u is the volume fraction where phase inversion is complete. Thus, polymers which have been mixed
while in a molten state have properties such as shown in
curve C of Figure 9. On the other hand, rigid fillers
which can not be fused together during mixing would have
properties which are characteristic of dispersed systems
and follow either curve A or curve B because such systems
can not undergo a phase inversion in which both phases
are continuous.

Models often have been used to predict properties
such as the elastic moduli or the permeability to liquids

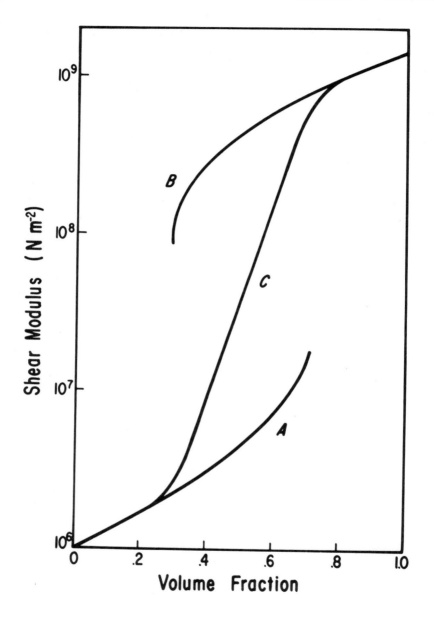

Figure 9.   Shear modulus as a function of concentration
for mixtures of two polymers and for filled polymers.
A. Rigid filler dispersed in a rubber.   B. Rubbery
filler dispersed in a rigid polymer.   C. Mixture of
two polymers which can undergo phase inversion. ϕ_m
= 0.74.   Equations 2, 9, and 10 apply between ϕ_2 = 0.26
and ϕ_2 = 0.74.

and gases of two-phase systems [9-12]. One example of
such models will be used to show how drastic changes in
morphology may have only slight effect on the properties
of a mixture which has two continuous phases [12].

Figure 10 shows two models. The continuous phases
made up of materials A and B are $A_{||}$ and $B_{||}$, respec-
tively. Part of each material may not be in a continuous phase, and this dispersed part is represented by $A_\perp$
and $B_\perp$. The volume fraction of the various parts are
proportional to the area of each of the rectangles making
up the model. The applied field (mechanical, electrical,
thermal or magnetic) acts in the direction indicated by
the arrows on the model. The models shown in Figure 10
are combinations of the simple models illustrated in
Figure 1 of Chapter 1. By combining the models for the
"rule of mixtures" and the "inverted rule of mixtures,"
it can be shown that the equations which apply to the
type of models shown in Figure 10 are [12]:

$$P = P_A \phi_{A||} + P_B \phi_{B||} + \frac{P_A P_B \left(\phi_{A\perp} + \phi_{B\perp} \right)}{P_A \phi_{B\perp} + P_B \phi_{A\perp}} \tag{11}$$

and

$$\phi_A = \phi_{A||} + \phi_{A\perp} \; ; \; \phi_B = \phi_{B||} + \phi_{B\perp} \; ; \; \phi_A + \phi_B = 1. \tag{12}$$

These models can be made to fit many types of experimental data. Generally, the values of the volume frac-
tions such as $\phi_{A||}$ and $\phi_{A\perp}$ are empirical, and it is
difficult to attach a physical significance to the
values of these parameters.

The two models in Figure 10 differ primarily in the
morphology of the soft phase A. In these models,
$\phi_A = \phi_B = 0.5$, and $\phi_{B||} = 0.3$, but the morphology as

LAWRENCE E. NIELSEN

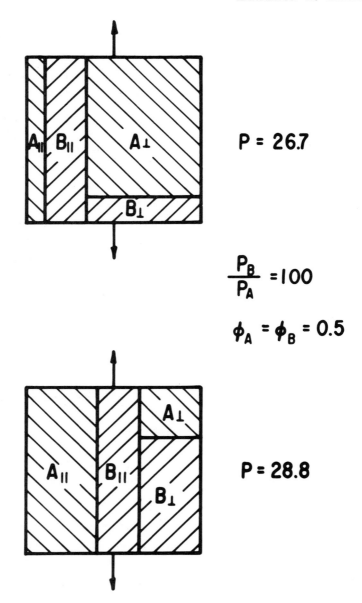

$$P = 26.7$$

$$\frac{P_B}{P_A} = 100$$

$$\phi_A = \phi_B = 0.5$$

$$P = 28.8$$

Figure 10.   Two models for which Equations 11 and 12 apply. $\phi_A = \phi_B = 0.5$. $\phi_{B\|} = 0.3$.   When $P_B/P_A = 100$, changing the nature of phase A has only a small effect on property P.

represented by $\phi_{A||}/\phi_{A\perp}$ changed from 1/4 to 4.0 in the
the two cases. When $P_B/P_A = 100$, the predicted value of
the property P only changed from 26.7 to 28.8 in the two
models. Thus, in some cases, drastic changes in morphology have little effect on the properties.

The connectivity of a phase is the fraction of it
which behaves as though it is a continuous phase. Thus,
the connectivity C_A of phase A is defined as

$$C_A = \phi_{A||}/\phi_A \tag{13}$$

while the connectivity C_B of phase B is

$$C_B = \phi_{B||}/\phi_B. \tag{14}$$

If the models are going to fit an equation such as Equation 1, then

$$\phi_{B||} = \frac{\left(P_B^n \phi_B + P_A^n \phi_A\right)^{1/n}}{P_B} - \frac{P_A}{P_B}\phi_{A||} - \frac{P_A\left(\phi_{A\perp} + \phi_{B\perp}\right)}{P_A \phi_{B\perp} + P_B \phi_{A\perp}}. \tag{15}$$

In such cases, Equation 15 indicates that the connectivity
of a phase approaches the total amount of that material
when its volume fraction is near one, but the connectivity
becomes very small as the volume fraction of a material
becomes less than about 0.2. This is to be expected
since a material tends to be a dispersed phase when it
has a low concentration.

In summary, although equations derived from complex
physical models can be useful, unless there is some theoretical justification for their use, it is generally best
to use Equation 1 or some other equation which has been
described in one of the previous chapters.

IV. REFERENCES

1. J. R. Partington, Advanced Treatise in Physical
    Chemistry, Vol. 2, Longmans Green, London, 1951,
    pp. 117.

2. A. Y. Coran and R. Patel, J. Appl. Polym. Sci., 20,
    3005 (1976).

3. W. E. A. Davies, J. Phys., D4, 318 (1971).

4. W. E. A. Davies, J. Phys., D4, 1176, 1325 (1971).

5. G. Allen, M. J. Bowden, S. M. Todd, D. J. Blundell,
    G. M. Jeffs, and W. E. A. Davies, Polymer, 15, 28
    (1974).

6. L. E. Nielsen, J. Appl. Polym. Sci., 19, 1485 (1975).

7. L. E. Nielsen, Rheol. Acta, 13, 86 (1974).

8. L. E. Nielsen, Mechanical Properties of Polymers
    and Composites, Vol. 2, Marcel Dekker, New York,
    1974.

9. M. Takayanagi, Proc. 4th Inter. Congr. Rheology, 1,
    161 (1965).

10. D. Prevorsek and R. H. Butler, Inter. J. Polymer
    Mater., 1, 251 (1972).

11. L. E. Nielsen, J. Macromol. Sci. (Chem.), A1, 929
    (1967).

12. L. E. Nielsen, J. Appl. Polym. Sci., 21, - (1977).

AUTHOR INDEX

Numbers in parentheses are reference numbers and indicate
that an author's work is referred to although his name is
not cited in the text. Underlined numbers give the page
on which the complete reference is listed.

A

Allen, G., 9(10), <u>19</u>, 83,
    85(5), <u>92</u>
Ashton, J. E., 50, 51, 52,
    57(12), <u>71</u>

B

Badger, W. L., 41(8), <u>47</u>
Blundell, D. J., 9(10), <u>19</u>,
    83, 85(5), <u>92</u>
Bowden, M. J., 9(10), <u>19</u>,
    83, 85(5), <u>92</u>
Brenner, H., 50, 52(3), <u>70</u>
Bruggeman, D. A. G., 51, <u>63</u>
    (14), <u>71</u>
Budiansky, B., 51(22), <u>71</u>
Burgers, J. M., 50, 57(4),
    <u>70</u>
Butler, R. H., 89(10), <u>92</u>

C

Carmichael, L. T., 14(12),
    <u>19</u>
Cheng, S. C., 51(25), <u>71</u>
Cigna, G., 67, 68(44), <u>72</u>
Coran, A. Y., 82(2), <u>92</u>

D

Davies, W. E. A., 9(9),
    9(10), <u>19</u>, 83, 85(3),
    83(4), <u>83</u>, 85(5), <u>92</u>
Di Benedetto, A. T., 50, 51
    (10), <u>70</u>

E

Einstein, A., 57(39), <u>72</u>

F

Frisch, H. L., 50, 52, 57(5),
    <u>70</u>

G

Gordon, M., 23(4), <u>47</u>, 62(42),
    <u>72</u>
Grieves, R. B., 35(6), <u>47</u>

H

Halpin, J. C., 8(7), <u>19</u>, 50,
    51, 52, 57(12), <u>71</u>
Hammond, L. W., 38(7), <u>47</u>
Happel, J., 50, 52(3), <u>70</u>
Hashin, Z., 51(19), 51, <u>60</u>
    (20), <u>71</u>, 60(41), <u>72</u>
Heusch, R., 43(10), <u>48</u>
Hildebrand, J. H., 22(3), <u>47</u>
Hirooka, M., 46(12), <u>48</u>
Hoftyzer, P. J., 62(43), <u>72</u>
Howard, K. S., 38(7), <u>47</u>

I

Illers, K. H., 43(11), <u>48</u>

J

Jeffs, G. M., 9(10), <u>19</u>, 83,
    85(5), <u>92</u>
Jenckel, E., 43(10), <u>48</u>

K

Kardos, J. L., 50, 51(10), <u>70</u>
Kato, T., 46(12), <u>48</u>
Kerner, E. H., 8(6), <u>18</u>, 51,
    60(18), 51(24), <u>71</u>
Klute, C. H., 51(30), <u>71</u>

SUBJECT INDEX

A

Adhesion between phases, 50

B

Block polymers, 73, 85

C

Classes of mixtures, 2
Concentration variables, 11
Connectivity, 74, 91
Copolymers, T_g of, 43
Critical temperatures, 35
Crystalline polymers,
    modulus of, 73, 83

D

Definition of concentration
    variables, 11
Density of mixtures, 38
Dielectric constant of
    liquid suspensions, 63
Dispersed systems, 49
Distorted concentration
    scales, 15, 35

E

Einstein coefficients, 52,
    57, 58

G

General mixture rules, 6
Glass transition tempera-
    tures, 43, 62

I

Interaction terms, 8, 21,
    23, 26
Interpenetrating networks,
    73

Invalid use of mixture rules,
    16
"Inverse rule of mixtures,"
    5, 10, 76
Inverted systems, 67, 85

L

Laminates, 73
Liquid-vapor phase diagrams,
    41
Logarithmic rule of mixtures,
    10, 75, 87

M

Maximum packing fraction, 8,
    49, 52, 53, 88
Mixtures, classes of 2
Mixtures, one-phase, 21, 74
Mixtures, two-phases, 49, 73
Mixtures with two continu-
    ous phases, 73
Models, 87
Mole fraction, definition,
    14

O

One-phase mixtures, 21, 74
One-phase mixture equations,
    8, 22, 23

P

Permeability of filled
    plastics, 69, 87
Plasticized polystyrene,
    T_g of, 43
Polyblends, 73, 85
Practical examples, one-phase
    mixtures, 35
Practical examples, two-phase
    mixtures, 62, 83